Principles of Chemical Nomenclature
A Guide to IUPAC Recommendations
2011 Edition

Principles of Chemical Nomenclature
A Guide to IUPAC Recommendations
2011 Edition

Edited by

G. J. Leigh
University of Sussex, Brighton, UK

IUPAC

RSCPublishing

ISBN: 978-1-84973-007-5

A catalogue record for this book is available from the British Library

Published by The Royal Society of Chemistry,
Thomas Graham House, Science Park, Milton Road,
Cambridge CB4 0WF, UK

Registered Charity Number 207890

For further information see our web site at www.rsc.org

Printed and bound in the United States of America.

Preface

The first version of this book was written in the conviction that although precise and accurate nomenclature is a vital aspect of professional training in the discipline of chemistry, few teachers, and fewer students, felt the need to master it. It has often been possible to convey information by using customary short and familiar names. However, that is as if the familiar and affectionate names we use for our families and friends were adequate for preparing a list of all the voters on a registration list. For more serious communication purposes, and above all for legal and regulatory reasons, proper names and details are required. Exactly the same requirements apply to chemical names and nomenclature. The purpose of this book, as it was of its predecessor, is to enable teachers and students up to the level required of university students in the early years of their studies to obtain a sound training in IUPAC chemical nomenclature.

How far we were successful in achieving this end with the first version is not completely clear. Some thought we had done very well, others thought we were still too detailed for elementary students. For that reason we have tried a different approach to writing this version, though still providing a clear text with many examples. This is because most students will scan the text until they find an example that seems to be of the kind they wish to copy. However, we have also added discussions of other nomenclatures that students may encounter, though we do not cover them in depth. We have also added information concerning current nomenclature developments that will become more significant in future, such as the IUPAC International Chemical Identifier (InChI) and the Preferred IUPAC Names (PINs). We have decided not to lower the overall standard, in the hope that those requiring only the more elementary material will be able to find what they want without undue effort. Finally, we wished to detail changes in IUPAC recommendations since the first edition of this book was written in 1997.

This expansion of coverage has required more expertise than we had available to the editor and writers (H. A. Favre, G. J. Leigh, and W. V. Metanomski) of the first version. Although there is still an overall editor (G. J. Leigh), the content was divided amongst a group of different writers, in alphabetical order: M. A. Beckett, Bangor University, UK; J. Brecher, CambridgeSoft, Cambridge, USA; T. Damhus, Novozymes, Copenhagen, Denmark; H. A. Favre, University of Montréal, Canada; R. M. Hartshorn, University of Canterbury, New Zealand; S. R. Heller, National Institute of Standards, Gaithersburg, USA; K.-H. Hellwich, Offenbach, Germany; M. Hess, Siegen

Principles of Chemical Nomenclature: A Guide to IUPAC Recommendations
Edited by G. J. Leigh
© International Union of Pure and Applied Chemistry 2011
Published by the Royal Society of Chemistry, www.rsc.org

University, Germany, and Chosun University, Republic of Korea; A. T. Hutton, University of Cape Town, South Africa; G. J. Leigh, University of Sussex, UK; A. D. McNaught, Royal Society of Chemistry, Cambridge, UK; G. P. Moss, Queen Mary University of London, UK. The writers were assisted by selected reviewers to look at the individual contributions before final editing to ensure accuracy and clarity. These were as follows, in alphabetical order: J. Brecher, CambridgeSoft, Cambridge, USA; N. G. Connelly, University of Bristol, UK; H. B. F. Dixon, University of Cambridge, UK; K.-H. Hellwich, Offenbach, Germany; R. G. Jones, University of Kent, UK; J. Kahovec, Academy of Sciences of the Czech Republic, Prague; W. V. Metanomski, Chemical Abstracts Service, USA; E. Nordlander, University of Lund, Sweden; J. Nyitrai, Budapest University of Technology and Economics, Hungary; W. H. Powell, Chemical Abstracts Service, USA. In addition, four meetings of the whole group, in Turin, Italy, in Büdingen, Germany, in Cambridge, UK and in Glasgow, UK, were convened in an attempt to ensure consistency and uniformity of the complete version. The editor here expresses his thanks to all those involved for their good humour, patience, hard work, and use of their expertise in the preparation of the manuscript. It was no easy matter to marshal such a large group of expert but argumentative and sometimes pedantic individuals. They all responded constructively and amicably to his sometimes brusque and uninformed editorial efforts. The editor would particularly like to acknowledge Karl-Heinz Hellwich for his meticulous proofreading.

Here it is in order to mention our common sadness at the passing of two persons who have been involved in the preparation of various stages in both this and the earlier version. Val Metanomski was the author of the Polymer Nomenclature chapter in the first edition of the *Principles of Chemical Nomenclature* as well as a reviewer of the current version, and was a respected and honoured member of the Commission on Macromolecular Nomenclature, then the Subcommittee on Polymer Terminology and of other IUPAC bodies including the Polymer Division (Division IV), the Interdivisional Committee on Terminology, Nomenclature and Symbols, ICTNS, and its predecessor, the Interdivisional Committee on Nomenclature and Symbols, IDCNS, for more than 20 years. Hal Dixon reviewed the Biochemical Nomenclature contribution to this version, and was a former secretary and chairman of the IUBMB-IUPAC Joint Commission on Biochemical Nomenclature and the Nomenclature Committee of IUBMB. He was involved with the work of these two bodies from their inception and made many valued contributions to biochemical nomenclature.

Nomenclature, just like chemistry, is a subject that develops continually. New classes of compound require new adaptations of nomenclature and IUPAC tries to provide these. For that reason, this version of *Principles* is unlikely to be the last of its kind. However, as the basic principles of nomenclature are unlikely to change, its content should continue to be of use. Like all the IUPAC Recommendations, it is the result of the efforts of an international group of experts, whose work and time is freely given. It represents the product of their endeavours over many years. However, any further version of *Principles* would be immeasurably improved if IUPAC were to receive copious comments and feedback from those to whom this version is directed. Please do not hesitate to address your reactions, however critical (or complimentary) to IUPAC at secretariat@iupac.org, stating clearly that your comments concern this latest version. You will be doing the next generation of chemists and nomenclature experts a considerable service.

Please note that, like all IUPAC recommendations and documents, this book is written in English, and the spellings and grammar employed here are British English usages.

G. J. Leigh
Sussex, UK

Contents

Principles of Chemical Nomenclature: A Guide to IUPAC Recommendations
Edited by G. J. Leigh
© International Union of Pure and Applied Chemistry 2011
Published by the Royal Society of Chemistry, www.rsc.org

CONTENTS

CONTENTS

1   Introduction

Chemical nomenclature is at least as old as the pseudoscience of alchemy, which was able to recognise a limited number of materials that could be reliably and reproducibly prepared. These were assigned names such as blue vitriol, oil of vitriol, butter of lead, aqua fortis, *etc.* that generally conveyed something of the nature of the material. Often these names were intended to be understood only by initiates, so they comprised a kind of private language. As chemistry became a real science, and the principles of the modern atomic theory and chemical combination and constitution were developed, such names no longer sufficed for general communication, and the necessity of developing systematic nomenclatures was realised by the end of the eighteenth century. Guyton de Morveau, Lavoisier, Berthollet, Fourcroy and Berzelius are amongst those notable for early contributions.

Nowadays even simple compounds can carry a host of names, some of obvious meaning, but others, especially trade names, are not. An inorganic example is sulfuric acid, which can be found in the chemistry literature, including scientific papers and indexes of various kinds, under a host of names. Names often change with the language employed, so IUPAC has decided to write its recommendations in English, leaving others to adapt them to different languages. Trade designations often are of the form 00647_FLUKA, and 35354_RIEDEL, and these can be solved using the appropriate trade catalogue. Other designations are simply numbers, and then the generator of the number should be consulted to discover the meaning. Chemical names or formulae in English texts for sulfuric acid, not all of them accurate, include: acidum sulfuricum; battery acid; BOV; dihydrogen sulfate; dihydrogen tetraoxosulfate; dihydroxidodioxidosulfur; dipping acid; electrolyte acid; H_2O_4S; H_2SO_4; hydrogen sulfate; hydrogen tetraoxosulfate(2−); hydrogen tetraoxosulfate(VI); Matting acid; methanolic H_2SO_4; oil of vitreol; oil of vitriol; OLEUM; SFO; $[SO_2(OH)_2]$; SO_3; SO_4; SOH; $[S(OH)_2O_2]$; sulfate; sulfate standard concentrate 10.00 g SO_4^{2-}; -sulfuric acid; sulfuric acid [acid aerosols including mists, vapors, gas, fog and other airborne forms of any particle size]; sulfuric acid, fuming; Sulfuric acid (NF); sulfuric acid solution; sulfuric acid, spent; sulfuric acid, spent [UN1832] [Corrosive]; sulfuric acid standard solution; sulfuric acid [Strong inorganic acid mists containing sulfuric acid]; sulfuric acid with >51% acid [UN1830] [Corrosive]; sulfuric acid with more than 51% acid [UN1830] [Corrosive]; sulfuric acid with less than 51% acid [UN2796] [Corrosive]; Sulfur oxide (SO_4); sulphuric acid; tetraoxosulfuric acid; Vitriol Brown Oil. Some of these names refer to mixtures rather than simple compounds.

The organic compound toluene has a similar set of trade names and related synonyms in English language publications. Names in trade indexes and other names and designations include the following terms: antisal 1a; benzene, methyl;

Principles of Chemical Nomenclature: A Guide to IUPAC Recommendations
Edited by G. J. Leigh
© International Union of Pure and Applied Chemistry 2011
Published by the Royal Society of Chemistry, www.rsc.org

benzene, methyl-; MBN; methacide; methane, phenyl-; methylbenzene; methyl-Benzene; methylbenzol; methylene, phenyl-; monomethyl benzene; otoline; phenyl methane; phenylmethane; phenylmethylene; TOL; toluene; toluene-(ring-UL-14C); toluene (Technical); toluene [UN1294] [Flammable liquid].

Clearly a student or a non-technical person may be overwhelmed by such a plethora of names. One function of IUPAC is to provide an unequivocal name based upon chemical structure so that all authorities can agree on what compound they are talking about. This book is an attempt to provide a teacher and student guide to a rather complicated subject.

The growth of chemistry in the nineteenth century was associated with the development of broader and more systematic nomenclatures, and chemists such as Liebig, Dumas, and Werner are associated with these innovations. During that century it was soon recognised that a systematic and internationally acceptable system of organic nomenclature was necessary. In 1892 the leading organic chemists of the day gathered in Geneva to establish just such a system. The Geneva Convention that they drew up was successful only in part. However, it was the forerunner of the current activities of the International Union of Pure and Applied Chemistry (IUPAC) and its Chemical Nomenclature and Structure Representation Division, within the Union known familiarly as Division VIII. Today this Division has the remit which until recently was devolved to the Commission on the Nomenclature of Organic Chemistry (CNOC), to the Commission on the Nomenclature of Inorganic Chemistry (CNIC), and to the Commission on Macromolecular Nomenclature. These Commissions have now been wound up, but their duties, to study all aspects of the nomenclature of chemical substances, to recognise and codify trivial (*i.e.*, non-systematic) nomenclature as appropriate, and to propose and recommend new systematic nomenclature as required by developments in chemistry, are now undertaken by Division VIII. This reorganisation has enabled the Union to treat more easily areas of nomenclature where more than one Commission had interests, sometimes conflicting. Organometallic chemistry is a case in point. The reorganisation has also facilitated collaboration between Division VIII and the Analytical Chemistry Division (V), the Inorganic Chemistry Division (II), the Organic and Biomolecular Chemistry Division (III), and the Polymer Division (IV), and also with bodies such as the International Union of Biochemistry and Molecular Biology (IUBMB). The recommendations described in this book are derived from all these bodies.

The systematic naming of substances, the presentation of formulae, and the drawing of structures involve construction of names and formulae from units that are manipulated in accordance with defined procedures in order to provide information on composition and structure. There are a number of accepted systems for doing this, of which the principal ones will be discussed below. In the text throughout this book, always remember that **bold type** is not a formalism used in IUPAC nomenclature, and its use here is solely to bring a given term to the special attention of the reader.

Whatever the pattern of nomenclature, names and formulae are constructed from units that fall into the following classes.

Element names, and **word roots** derived from these element names, and **element symbols**.

Parent molecule names, which are names used as bases of nomenclature, and which can be manipulated in various ways, as described below, to develop the names of formal derivatives of these molecules; the commonest kind of parent is a **parent hydride**.

Prefixes, which are defined in the English language as verbal elements placed before and joined to a word or stem to add to or qualify its meaning. Many prefixes consist of a set of letters or characters placed before a name to indicate atoms or groups of atoms which are regarded either as **substituents** (see Section 6.2.5) or **ligands** (see Section 7.2.3), and also to indicate aspects of stereo-chemistry (see Sections 3.8 and 7.5.7); **numerical prefixes** or **multiplicative prefixes** specify the number of groups of a specific kind that are present (see Table P6). When numbers (and less often other characters) are used to indicate the position of particular atoms or groups of atoms within a chemical structure, these are termed **locants**. Sometimes numbers may also be used to indicate particular aspects of stereochemistry (see Section 3.9).

Infixes, either numerical or alphabetical, which are usually locants inserted into a name and separated from the letters forming the name by hyphens (see Section 4.3), though infixes also have other functions.

Suffixes, sets of letters or characters, which are added at the end of a name to indicate properties of a substance such as charge, oxidation state and characteristic group.

Additive prefixes and/or suffixes, which are sets of letters or characters indicating the formal addition of particular atoms or groups to a parent molecule.

Subtractive prefixes and/or suffixes, which are sets of letters or characters indicating the absence of particular atoms or groups from a parent molecule.

Descriptors, which give information upon aspects such as the structure, geometry, and configuration of the substance being named.

Punctuation marks, including **commas, hyphens, colons, semicolons**, and **full stops** (or **periods**).

There may be a problem with hyphens in names because they are used in conventional typography to break a word at the end of a line when there is no space to finish that word, but they are also a fundamental symbol in nomen-clature. Unfortunately many chemical names are very long, and frequently they need to be broken at the end of a line of type. Using a hyphen to do this when hyphens are also integral parts of the name may often be confusing to the reader. Consequently, *in this text, and in this text alone*, we use the symbol ▶ to indicate a line break where one might use a hyphen in normal text. This symbol is not part of nomenclature or a name, and should never appear used as such in a printed text. It should be omitted in the general application of nomenclature rules. Its use here is solely to inform an inexperienced reader. In addition, it may confuse some readers when suffixes, endings, prefixes and infixes are specified in a text by examples such as cyclo-, -ate, -nitrile, and -1,2-. It may not be clear whether the hyphens are to be retained when such units are built into a name. To avoid this problem, all prefixes, suffixes, endings, and infixes are rendered in this text in the form 'cyclo', 'nitrile', and '1,2'. The reader will have to ensure that the correct use of hyphens is observed in a name derived using such elements.

The uses of all these nomenclature tools will be exemplified in the following Chapters. The material discussed is based on *Nomenclature of Organic Chemistry*, 1979 (sometimes referred to as the IUPAC Blue Book). This edition is out of date in places and a new version is expected to be published soon, but updated information as well as html versions of this and other IUPAC nomenclature publications can be found on the IUPAC website at http://www.iupac.org, and in *A Guide to IUPAC Nomenclature of Organic Compounds, Recommendations 1993*, that may be found to be more accessible, the *Nomenclature of Inorganic Chemistry, IUPAC Recommendations 2005* (also often referred to as the Red Book), the *Compendium of Polymer Terminology and Nomenclature, Recommendations* 2008 (the Purple Book), and *Biochemical Nomenclature and Related Documents* (the White Book) issued by IUBMB in 1992. Other useful sources are *Quantities, Units and Symbols in Physical Chemistry*, 2007 (the Green Book) and the *Compendium of Chemical Terminology, 1997* (the Gold Book, a later interactive version of which can be found at http://goldbook.iupac.org). These so-called colour books are the principal printed vehicles for IUPAC nomenclature recommendations, but most of them, and later recommendations, will generally also be found on-line.

For historical reasons more than one IUPAC nomenclature system has developed, and in many cases more than one IUPAC name can be suggested for a particular compound. Often a preferred name will be designated, but as there are several systematic or semi-systematic nomenclature systems in use it may not be easy or advisable to recommend an unique name. IUPAC is currently attempting to resolve this conundrum in a satisfactory fashion, by developing rules for selecting Preferred IUPAC Names (PINs). In addition, many non-systematic (trivial) names are still in general use. Although it is hoped that these will gradually disappear from the literature, many are still retained for current use, although often in restricted circumstances. These restrictions are described in the text. The user of nomenclature should adopt the name most appropriate for the purpose in hand.

Finally, it should be noted that IUPAC nomenclature seeks to be as systematic as possible, and to facilitate both the easy comprehension of names and their facile correct production. IUPAC nomenclature is preferred for legal and business transactions by international bodies, such as the European Union. However, other regulatory bodies have developed more limited nomenclature systems appropriate to their specific uses. Such systems include those for pharmaceuticals, international non-proprietary names (INNs), vitamins, fatty acids, enzymes, *etc*. It is not the intention to deal with these in any detail in this publication, though some are described briefly in Chapter 12, and Table P10 (for a brief discussion of Tables in this book, see below) lists some names from non-IUPAC systems that may be met with in day-to-day usage, together with their systematic IUPAC equivalents.

The Chemical Abstracts Service (CAS) Division of the American Chemical Society is a major developer and user of chemical nomenclature. The system used by CAS is often similar to IUPAC's, but not identical. It is necessary for the Chemical Abstracts Service to name compounds systematically as soon as they appear in the chemical literature, and the names must be recognisable by the computer systems which it employs. Generally CAS cannot afford to wait until the much slower deliberations of IUPAC have come to fruition. Consequently the two systems diverge, though IUPAC is generally accepted to provide the international standard. Readers of the chemical literature should be aware of the nomenclature system being used in any given publication.

IUPAC has also recently developed recommended methods for discussing and drawing representations of two- and three-dimensional chemical structures. These are summarised in Chapter 16, but will not be discussed in any detail. Anyone wishing to explore this matter further should consult the latest publications, currently Graphical Representation of Stereochemical Configuration (IUPAC Recommendations 2006), *Pure and Applied Chemistry*, 2006, **78**(10), 1897–1970, and Graphical Representation Standards for Chemical Structure Diagrams (IUPAC Recommendations 2008), *Pure and Applied Chemistry*, 2008, **80**(2), 277–410. The name of the journal, *Pure and Applied Chemistry*, is frequently abbreviated to *Pure Appl. Chem.*, and even represented by an acronym, PAC. Students will find most of this content too advanced, but teachers may find it helpful in inculcating good practice.

This book is provided with many cross-references. The usage adopted is that the first numeral in a Section or Table reference is the number of the Chapter to which it refers. Thus Section 5.7.2 refers to Section 7.2 of Chapter 5 and Table 8.7 is the seventh table of Chapter 8. Tables designated with the letter P, such as Table P7, are referred to in more than one Chapter, and these are gathered together at the back of the book. Table P7 and Table P10 contain lists of names that will be useful to a reader searching for group names and trivial names, for example to apply in new written text. Readers are recommended to become generally aware of the contents of all the P Tables.

Each Chapter contains its own set of references, numbered sequentially and listed at the end of that Chapter.

2 Definitions

ATOMS AND ELEMENTS

An **element** (or an elementary substance) is matter, the atoms of which are alike in having the same positive charge on the nucleus. This positive charge is termed the **atomic number**.

In certain languages, a clear distinction is made between the terms 'element' and 'elementary substance'. The term 'element' is one that is sometimes considered to be an abstraction. It implies the essential nature of an atom, and this is retained however the atom may be combined, or in whatever form it exists. An 'elementary substance' is a physical form of that element, as it may be prepared and studied. In the English language, it is not customary to make such a fine distinction, and the word 'atom' is sometimes used interchangeably with the words 'element' or 'elementary substance'. Particular care should be exercised in the use and comprehension of these terms.

An **atom** is the smallest unit quantity of an element that is capable of existence, whether alone or in chemical combination with other atoms of the same or other elements. The life-times of some kinds of atom may be essentially infinite, but other types of atom may decompose after only a fleeting existence. In the latter case, establishing that the atom has actually existed may be difficult and contentious.

The elements are assigned names, some of which have origins deep in past history, but other names are relatively modern. All these names are termed 'trivial' by nomenclaturists. The adjective 'trivial' in this context means that the names are not 'systematic', but are derived from a variety of sources, ancient and modern. The only element names that are systematic are those temporarily assigned to elements the existence of which may be discussed but which has yet to be unequivocally established.

The **element atomic symbols** consist of one, two, or three roman letters, often but not always, related to the name in English. In the examples below, the names in Examples 1–5 are trivial, but the name in Example 6 is systematic.

Examples
  1. hydrogen   H
  2. argon   Ar
  3. potassium   K
  4. sodium   Na
  5. chlorine   Cl
  6. ununseptium   Uus

Principles of Chemical Nomenclature: A Guide to IUPAC Recommendations
Edited by G. J. Leigh
© International Union of Pure and Applied Chemistry 2011
Published by the Royal Society of Chemistry, www.rsc.org

For a list of elements correct to July 2010, see Table P1. For the heavier elements as yet unnamed or not yet synthesised, the three-letter symbols, such as Uus and their associated names, are temporary. They are provided for use until such time as a consensus is reached within the chemistry and physics communities that these elements have indeed been discovered or synthesised, and a trivial name and symbol have been assigned after the prescribed IUPAC procedures have taken place.

2.2 THE PERIODIC TABLE AND ELEMENT SYMBOLS

When the elements are suitably arranged in order of their atomic numbers, a Periodic Table is generated. There are many variants, in shape and in the amount of information they are designed to convey, and chemists will use whatever kind is most useful to them. Elements fall into various classes, as laid out in a typical Periodic Table (Table P2). The vertical alignments of elements are referred to as **groups**, and the horizontal alignments are referred to as **periods**. For teaching purposes, some authorities may prefer extended or amended versions of the Table P2, to reflect particular aspects of properties such as electronic structure. However, IUPAC is concerned principally with rapid and unambiguous communication, and accessibility to all members of the chemistry community at every level of understanding. The 1–18 group numbering shown in Table P2 was adopted in 1985 to replace the contradictory designations that were then in vogue. Whatever form of the Periodic Table may be adopted by given individuals, it is recommended that the 1–18 group numbering should always be observed. Designation of periods tends to be less confusing, and terms such as **First Short Period** and **Second Long Period** are current and generally clear. IUPAC has not recommended names for periods. For groups see page 9.

An atomic symbol can carry up to four modifiers (some authorities prefer the term 'index' to 'modifier') to convey further information. This is shown below for a hypothetical atom, symbol X.

$$_{C}^{D}X_{B}^{A}$$

The modifier A indicates a **charge number**, which may be positive or negative, as in 2+ or 2–, when atom X is in a form more properly called an **ion**. Note that it is never correct to write charges in the forms +2 or –2, or ++ or ––. When the modifier A is not shown, the charge is assumed to be zero. Alternatively or additionally A can indicate **the number of unpaired electrons** on X, in which case the modifier is a combination of an arabic numeral and a dot, as in 2·. When charges or unpaired electrons are indicated, the number 1 is usually not represented. In Example 6 below, note that in the position of the modifier A, both the number of unpaired electrons and the charge are indicated. These should be written in the order illustrated here.

Examples
1. Na^+ 2. Cl^-
3. Ca^{2+} 4. O^{2-}
5. N^{3-} 6. $N^{\cdot 2-}$

Finally, the position of modifier A may also be used for a Roman numeral to indicate the oxidation state assigned to an atom when that atom is represented by its atomic symbol (not by its name). This formalism should not be used when the atomic charge and/or the number of unpaired electrons are also being specified in this position.

Modifier B indicates the number of atoms bound together in a single chemical entity or species. If the value of B is 1, it is not represented. In a formula such as an **empirical formula** (see Chapter 3), B can be used to indicate relative proportions.

Examples
7. P_4 8. S_8
9. Cl_2 10. C_{60}

Care must be exercised in the precise placing of modifiers A and B. It is evident that Hg_2 with the modifier B represents a pair of mercury atoms. However, if it is desired to show that the pair of mercury atoms carries a 2+ charge, caution is necessary. Some word processing programs do not easily allow the placing of superscripts immediately over subscripts, but this may not be, in any case, desirable because then it may not be clear whether the two mercury atoms each carry a charge of 2+, or the two mercury atoms carry a total charge of 2+. The solution is to offset either A or B. Reading the sequence of modifiers from left to right in the usual fashion, as in Hg^{2+}_2, clearly implies the presence of two mercury atoms each with a charge of 2+, which is not the intention here. An earlier recommendation, which can still be applied, is to use parentheses, *viz.*, $(Hg_2)^{2+}$, but the parentheses may easily be omitted giving Hg_2^{2+} without changing the meaning, as long as the sequence of modifiers is not changed. This presentation, with the right superscript offset, may be adapted to all similar situations, in particular to avoid ambiguity.

Modifier C is used to denote the **atomic number**, but this space is generally left empty because the atomic symbol necessarily implies the atomic number.

The modifier D is used to show the **mass number** of the atom being considered, this being the total number of neutrons and protons considered to be present in the nucleus. This is not the atomic weight. The number of protons defines the element, but the number of neutrons in the atoms of a given element may vary. Any atomic species defined by specific values of atomic number and mass number is termed a **nuclide**. Atoms of the same element but with different mass numbers are termed **isotopes**. This word means that they occupy the same position in the Periodic Table. The mass number can be used to designate specific isotopes.

Examples
11. ^{31}P
12. ^{1}H, ^{2}H (or D), ^{3}H (or T)
13. ^{12}C

Note that of all the isotopes of all the elements, only those of hydrogen, ^{2}H and ^{3}H, also have specific atomic symbols, D and T, with the associated names **deuterium** and **tritium**. The lightest isotope, ^{1}H, is sometimes referred to as **protium**. For all the elements, the unqualified atomic symbol is taken to represent a mixture of isotopes, though hydrogen represents a partial exception. The ions $^{1}H^{+}$, D^{+} and T^{+} are called **protons**, **deuterons**, and **tritons**, respectively. A positively charged hydrogen ion derived from a mixture containing all three forms may actually be any one of these three isotopes, so it is generally called by the non-specific term **hydron**.

An alternative way to represent specific isotopes is to quote the element name and the mass number, as shown below.

Examples
14. hydrogen-1   15. carbon-13
16. phosphorus-31   17. nitrogen-14

Elements are sometimes classified in a way to reflect some common properties. A generally recognised class is that of the **Main Group elements** (groups 1, 2, 13, 14, 15, 16, 17, and 18 of the Periodic Table), and the two elements of lowest atomic number in each group are often called **typical elements**. The elements of groups 3–12 are termed **transition elements**. The lightest element of all, hydrogen, may appear anomalous because it forms a class of its own in the Table illustrated. Some forms of the Periodic Table have been designed to avoid this anomaly. Other more trivial group designations, such as alkali metals, halogens, *etc.*, are in current use, but such names are not often used in nomenclature.

2.3   ELEMENTARY SUBSTANCES

Only a few elements form a monoatomic elementary substance. The majority form polyatomic materials, ranging from diatomic substances, such as H_2, N_2, and O_2, through polyatomic species, such as P_4 and S_8, to infinite three-dimensional polymers, such as the metals. Polyatomic species where the degree of aggregation can be precisely defined are more correctly termed **molecules**. However, the use of the word 'element' is not restricted to the consideration of elementary substances. Compounds are composed of numbers of atoms of one or more kinds in some type of chemical combination. For example, water is a compound of the elements hydrogen and oxygen. A molecule of water is composed of three atoms, two of which are of the element hydrogen and one of the element oxygen.

Molecules are often neutral, but they may become charged, positively or negatively, by various processes, when they are called **ions**. Charge is not common amongst elementary substances, but where some atoms or molecules are positively charged, these as a class are called **cations**. The term **molecular ion** is appropriate in some contexts such as mass spectrometry. For a substance to be overall neutral, these positively charged species must be accompanied by negatively charged species, and if these latter are atoms or molecules they are termed **anions**.

Many elements can give rise to more than one elementary substance. These may be substances containing assemblages of the same mono- or poly-atomic unit, but arranged differently in the solid state (as happens with tin), or they may be assemblages of different polyatomic units (as with sulfur and oxygen, and with carbon, which forms diamond, graphite, and the fullerenes). These different forms of an element which are composed of different polyatomic assemblages are referred to as **allotropes**. Their common nomenclature is essentially trivial, though attempts have been made to develop systematic nomenclatures, especially for crystalline materials. These attempts are not wholly satisfactory.

Throughout this discussion, we have been considering only pure substances, *i.e.*, substances composed of a single material, whether an element or a compound. A **compound** is a single chemical substance and it may be molecular, or ionic, or both. For example, sodium chloride is an ionic compound that contains two atomic species, Na^+ and Cl^-. If a sample of sodium chloride is somehow manipulated to remove some Cl^- ions and replace them by Br^- ions in equivalent number, the resulting material is a **mixture**. A substance containing various neutral species such as P_4, S_8, and C_6H_6 is also a mixture. In this context, a solution, consisting of a solute dispersed in a solvent, is also a mixture. There are, of course, an infinite number of possible mixtures of a large number of kinds, and IUPAC has not, with very few exceptions, developed systematic methodologies for naming them.

Pure substances (be they elementary or compound) and mixtures are usually solids, liquids, or gases, and they may even take some rarer form. These forms are termed **states of matter** and are not, strictly, within the province of nomenclature. However, to indicate by a name or a formula whether a substance is a solid, liquid or gas, the letters **s**, **l**, and **g** are used. For more details, see *Quantities, Units and Symbols in Physical Chemistry*, RSC Publishing, Cambridge, 2007 (the Green Book), and also the on-line version, which carries later information.

Examples
1. $H_2O(l)$
2. $H_2O(g)$
3. $H_2O(s)$

For further general definitions in chemistry, readers may wish to refer to the *Compendium of Chemical Terminology*, Blackwell Science, Oxford, 1997 (the Gold Book). An updated version can be obtained *via* http://goldbook. iupac.org. This and all IUPAC nomenclature publications are now made available on-line at http://www.iupac.org as soon as is feasible after final approval by the appropriate IUPAC bodies.

2.4 PREFERRED IUPAC NAMES

It will be evident after perusing the following chapters of this book that IUPAC often allows several different methods for naming a given compound. The user may adopt whichever is found to be most appropriate, but national and international regulatory and legal bodies generally require a single name to designate a specific compound, to enhance communication and to avoid confusion and misunderstanding. To this end, IUPAC, the international authority on nomenclature, is developing systems for selecting **Preferred IUPAC Names**, often called by the acronym **PINs**, from the various names recognised by IUPAC. These have been selected for many organic compounds. The work on the selection of PINs for inorganic compounds and polymers is less advanced than that for organic compounds, but an account of inorganic PINs is in preparation. Once the rules for selecting PINs have been codified, their general use throughout the chemistry community will be encouraged.

2.5 IUPAC INTERNATIONAL CHEMICAL IDENTIFIERS

Another system of nomenclature currently being developed by IUPAC is called the **IUPAC International Chemical Identifier**, or **InChI**. This is a machine-readable string of symbols which enables a computer to represent a compound in a completely unequivocal manner. InChIs are produced by computer from structures drawn on-screen, and the original structure can be regenerated from an InChI with appropriate software. An InChI is not directly intelligible to the normal reader, but InChIs will in time form the basis of an unequivocal and unique database of all chemical compounds. The necessary software will be made freely available. InChIs are already being used in some commercial databases, and for this reason, a brief discussion of InChIs is included in Chapter 14.

3   Formulae and Stereoformulae

3.1   INTRODUCTION

The basic materials of systematic chemical nomenclature are the element names and symbols, which, with the exception of the systematic temporary names and symbols for the elements with atomic number greater than 112, are, in nomenclature terms, trivial. These temporary names will be superseded eventually by permanent names and symbols. These heavy elements usually have very short half-lives, and are unlikely to be encountered often in general chemical practice.

The simplest way to represent chemical substances is to use formulae, which are assemblages of chemical symbols. Formulae are particularly useful for listing and indexing and also when names become long and complex. The selection of the precise form of a formula depends upon the use to which it is to be put.

3.2   EMPIRICAL FORMULAE

The simplest kind of formula of a compound is a **compositional** or **empirical** formula, which lists the constituent elements in the atomic proportions in which they are present. For such a formula to be useful in lists or indexes, an order of citation of symbols must be agreed first. Such an order, often termed a **seniority** list, is commonly used in nomenclature. Within a seniority list, higher terms take **priority** over lower. For lists of formulae and formulae indexes, the alphabetical order of atomic symbols is generally recommended. However, because both carbon and hydrogen are almost always present in organic compounds, the normal but not invariable practice in English is to cite C first, H second, and then the rest of the symbols in English alphabetical order. In compounds that do not contain carbon, strict alphabetical order should be adhered to in indexes.

Note that since IUPAC nomenclature rules are written in English, the alphabetical order corresponds to the English alphabet. Lists based upon languages other than English may be different, and may not be appropriate for international use.

Molecular or ionic masses cannot be calculated from empirical formulae, so that Example 2, below, could refer equally to several species, including, CH_2 itself, C_2H_4, and C_3H_6, and so on. Note also that in general practice this alphabetical order exemplified in Examples 1–9 may not be observed. For instance, Example 1 is usually written KCl, following the order based upon relative electronegativities (see below, Section 3.5).

Principles of Chemical Nomenclature: A Guide to IUPAC Recommendations
Edited by G. J. Leigh
© International Union of Pure and Applied Chemistry 2011
Published by the Royal Society of Chemistry, www.rsc.org

Examples

1. ClK
2. CH_2
3. CaO_4S
4. CH_2O
5. $C_6FeK_4N_6$
6. C_2Be
7. NS
8. $C_{10}H_{10}ClFe$
9. Cl_4Uuq

The chemistry of the heavy, unstable element Uuq is almost completely unexplored at present, so that the formulation in Example 9 is inferred from its presumed position in the Periodic Table, and is not based upon empirical observation. For further examples of seniority and priority used in organic and inorganic chemistry nomenclatures, see Table 3.2, Table 6.1, Table 6.2, and Chapters 6, 7 and 8.

3.3   MOLECULAR FORMULAE

Molecular formulae for compounds consisting of discrete molecules are formulae written according with the relative molar mass for the structure, sometimes also called the relative molecular mass or the molecular weight. Note that the ordering of symbols in these formulae below sometimes departs from the alphabetical (Example 2). The ordering principles employed here are discussed below (Section 3.5).

Examples

1. N_4S_4
2. S_2Cl_2
3. C_2H_6
4. C_{60}

Polyatomic ions are treated similarly to these polyatomic neutral molecules, although the charge must also be indicated. Such formulae imply nothing about chemical structure. As soon as structural information is combined with a formula, these simple rules need to be modified.

The discussion so far has assumed that all compounds are stoichiometric, that is, the atomic or molecular proportions are integral. It has become increasingly clear that many compounds are to some degree non-stoichiometric. The rules cited so far fail for non-stoichiometric compounds, for which further formalisms need to be developed. Of course, electroneutrality must be maintained overall in some way in such compounds. For example, in an ionic compound of the idealised formula MCl_x, where there is apparently a deficit of negative ions, the consequent formal excess of cations may be neutralised by the presence of an appropriate number of cations of the form $M^{(n-1)+}$ rather than the prevalent form M^{n+}, for example, as Fe^{2+} rather than Fe^{3+}. Various stratagems have been used to represent this kind of situation in formulae, although not yet in names. For further details, the reader is referred to the *Nomenclature of Inorganic Chemistry, IUPAC Recommendations 2005*, Chapter 11.

Examples

5. $\sim FeS$
6. $Co_{1-x}O$
7. $(Li_2,Mg)Cl_2$
8. $Fe_{1.05}Li_{3.65}Ti_{1.30}O_6$

3.4 STRUCTURAL FORMULAE

Structural formulae give information about the way atoms in a molecule or ion are connected and arranged in space.

Examples

1.

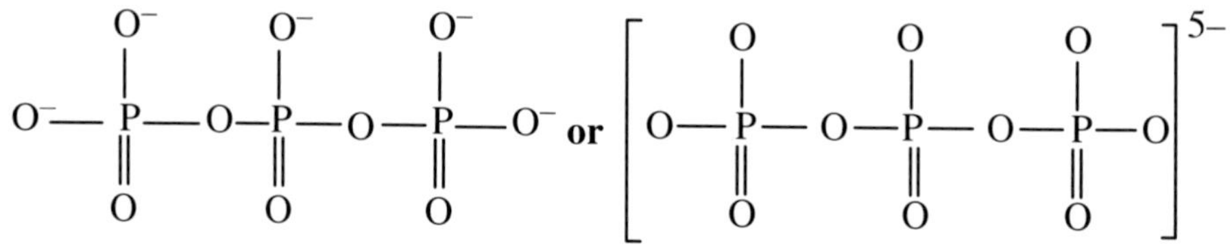

2.

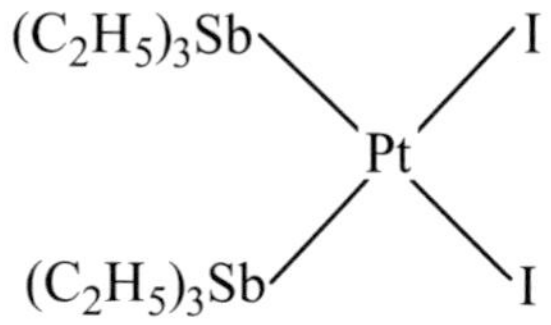

Attempts may be made to represent a structure in three dimensions.

Example

3.

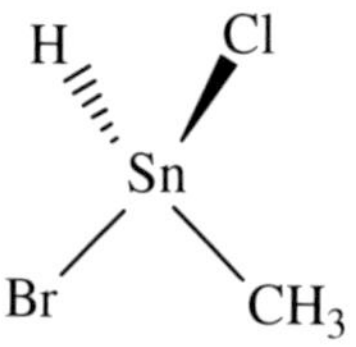

In this example, the full lines (——) represent bonds in the plane of the paper, the hashed wedge (⫶⫶⫶) represents a bond from tin pointing away from the viewer and below the plane of the paper, and the full wedge (——◗) a bond pointing towards the viewer and above the plane of the paper. This kind of representation will be discussed in more detail in Section 3.8. The use of dotted lines to represent bonds out of the plane of the paper, still used in some quarters, is no longer recommended. A complete account of IUPAC recommendations on this subject is to be found in reference 1.

In organic chemistry, structural formulae are frequently presented as condensed formulae. This abbreviated presentation is especially useful for large molecules. Another way of presenting structural formulae is by using bonds only, with the understanding that carbon and hydrogen atoms are never explicitly shown, except sometimes for terminal groups.

Examples

4.

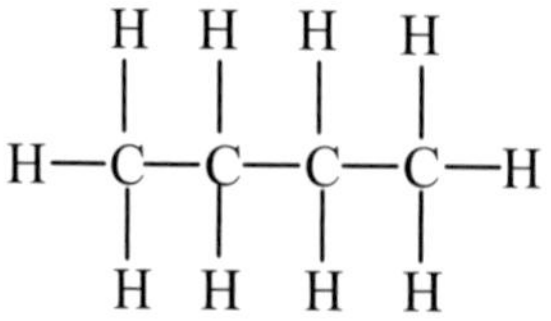

H—C—C—C—C—H or H_3C—CH_2—CH_2—CH_3

or $H_3CCH_2CH_2CH_3$ or ⌵⌃

5. $H_3C-CH_2-CH_2-CH_2-CH_2-CH_3$

or $CH_3CH_2CH_2CH_2CH_2CH_3$ or ⟋⟍⟋⟍⟋

6.

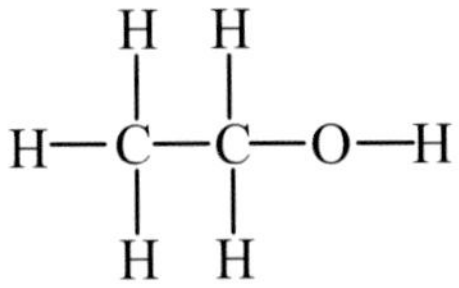

```
      H  H
      |  |
 H—C—C—O—H      or      H3C—CH2—OH
      |  |
      H  H
```

or CH_3CH_2OH or ⟍⟋OH

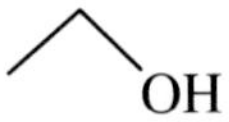

7.

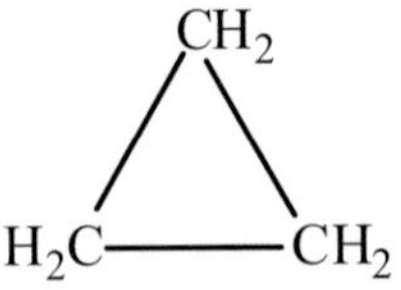

```
          CH2
         /    \
        /      \         or      △
   H2C————CH2
```

8.

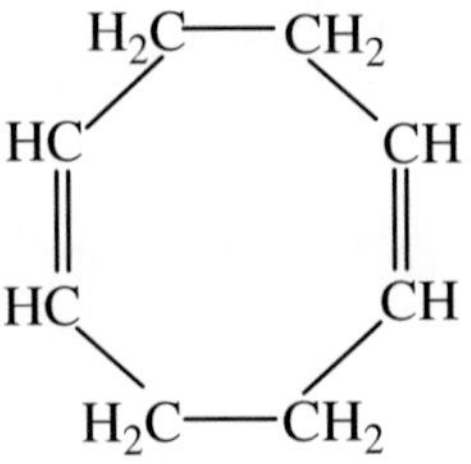

```
       H2C——CH2
      /          \
    HC            CH
    ‖             ‖        or      ⬡
    HC            CH
      \          /
       H2C——CH2
```

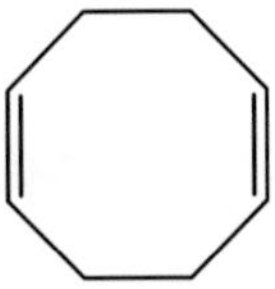

As will be evident from the above examples, and by extrapolation from the rules cited in Section 3.3, the numbers of groups of atoms in a unit and the charge on a unit are indicated by modifiers in the form of subscripts and superscripts.

Examples
   9. $C(CH_3)_4$
   10. $H_3C[CH_2]_5CH_3$
   11. $CaCl_2$
   12. $[\{Fe(CO)_3\}_3(CO)_2]^{2-}$

Note the use of enclosing marks: parentheses ( ), square brackets [ ], and braces { }. They are used to avoid ambiguity. In the specific cases of coordination compounds, square brackets denote a 'coordination entity' (see Section 7.4.6). In the organic Example 10 above, the use of square brackets to indicate an unbranched chain is shown. Note that in organic nomenclature generally and in inorganic names (but not formulae), the same classes of enclosing mark are used, ( ), { }, and [ ], in the sequence {[({[( )]})]}.

3.5   SEQUENCE OF CITATION OF SYMBOLS

As stated above, the sequence of atomic symbols in a molecular formula is arbitrary, but in the absence of any other requirements a modified alphabetical sequence is recommended. This is primarily a sequence for use in indexes, such as in a book.

Where there are no overriding requirements, the following criteria may be generally adopted. In a formula, the order of citation of symbols is based upon relative electronegativities. Although there is no general confusion about which of, say, Na and Cl is the more electronegative element, there is no universal scale of electronegativity that is appropriate for all purposes. However, for ionic compounds, cations are always cited before anions. In general, the decision as to which of two elements is the more electronegative may not be so evident. Therefore, some twenty years ago it was decided to use the sequence shown in Table P3 to represent an electronegativity scale for nomenclature purposes. The order of citation proposed in a binary compound is from the least electronegative (*i.e.*, the most electropositive) element to the most electronegative, and the least electronegative element is that encountered last on proceeding through Table P3 in the direction of the arrows. This is not to be taken to indicate any specific values for the electronegativity of an element, and no neutral point is delineated.

If a formula contains more than one element of each class, the order of citation within each class is alphabetical. Note, however, that a hydron in the form of 'acid hydrogen' is always regarded as an electropositive element, and immediately precedes the anionic constituents in the formulae of acids. Compare the practice with the names, as discussed in Section 5.6.1.

Examples

1. KCl
2. $Na_2B_4O_7$
3. $IBrCl_2$
4. O_2ClF_3
5. $NaHSO_4$
6. NaUus

Where it is known that certain atoms in a molecular ion are bound together to form a group, as with S and O in SO_4^{2-}, these elements can be so grouped in the formula, with or without enclosing marks, depending upon the compound and the requirements of the user.

Examples

7. HBr
8. H_2SO_4
9. $[Cr(OH_2)_6]Cl_3$
10. $H[AuCl_4]$

There are various sub-rules: for example, a single-letter symbol such as B always precedes a two-letter symbol such as Be; NH_4 is treated as if it were a two-letter element symbol and where appropriate would be listed after Ne. The written alphabetical ordering of a polyatomic group is determined by the first symbol cited: SO_4^{2-} by S; $[Zn(OH_2)_6]^{2+}$ by Zn; NO_3^- by N, *etc.* A more detailed discussion is given in the *Nomenclature of Inorganic Chemistry, IUPAC Recommendations 2005*, RSC Publishing, Chapter 4.

For binary compounds between non-metals (*i.e.*, between elements that are both usually considered to be electronegative), the electronegativity sequence defined by Table P3 is adopted, and the least electronegative element is cited first. The sequence of increasing electronegativity now used is shown here.

Rn Xe Kr Ar Ne He B Si C Sb As P N H Te Se S O At I Br Cl F

For intermetallic compounds, where all the elements can be considered to be electropositive, strict alphabetical ordering of symbols is recommended. However, the term 'intermetallic compound' is yet to be defined in a completely satisfactory manner, so that some flexibility is allowed, and other ordering principles, such as the use of Table P3, may be considered appropriate. The situation is summarised in the *Nomenclature of Inorganic Chemistry, IUPAC Recommendations 2005*, Sections IR-4.4.2 and IR-4.4.3.

Examples

   11. Au_2Bi 12. NiSn

3.6   **FORMULAE OF GROUPS**

We have already mentioned the formula for groups, such as $SO_4{}^{2-}$, without discussing the principles by which such formulae are assembled. These may (or may not) involve some reference to structure. The general approach is to select one or more atom(s) as the central or characteristic atom(s). This is so whether or not the ion or group is regarded as a coordination entity. Thus I in $ICl_4{}^{-}$, V in $VO_2{}^{+}$, and Si and W in $[SiW_{12}O_{40}]^{4-}$ are all central atoms and are cited before the ligating atoms. The subsidiary atoms then follow in alphabetical order of symbols, though this rule is slightly modified for coordination compounds.

Examples

   1. $[Cr_2O_7S]^{2-}$ 2. H_3PO_4
   3. $[ICl_4]^{-}$ 4. $SbCl_2F$
   5. OCl^{-} 6. $PBrCl_2$
   7. $NO_2{}^{-}$

Coordination compounds, which may be charged or uncharged are considered to be composed of a central atom to which the coordinated ligands donate electrons. The symbol of the central atom is cited first. The ligands in formulae of coordination entities are ordered alphabetically, according to the abbreviation or formula used for the ligand, irrespective of the formal charge. This is a change from the practice recommended previously (see *Nomenclature of Inorganic Chemistry, IUPAC Recommendations 2005*, Section IR-1.6.5). The formula of the whole coordination entity (which may be positive, negative, or neutral) is enclosed in square brackets. Because square brackets are always of highest priority in coordination formulae, a hierarchical sequence of enclosing marks is adopted to ensure that this seniority is preserved: [ ], [( )], [{( )}], [({( )})], [{({( )})}], *etc.* Note that the priority sequence used in both coordination names and organic substitutive names, ( ), [( )], {[( )]}, ({[( )]}), *etc.*, is different from that used in formulae.

Examples

   8. $[Ir(C_5H_5N)Cl_2H(NH_3)]$ 9. $K_3[Fe(CN)_6]$
  10. $[Ru(NH_3)_5(N_2)]Cl_2$ 11. $K_2[Cr(CN)_2(NH_3)O_2(O_2)]$
  12. $[Cu\{OC(NH_2)_2\}_2Cl_2]$ 13. $[ICl_4]^{-}$

It is sometimes a matter of individual preference whether a species is regarded as a coordination entity or not. For example, sulfate may be regarded as a coordination complex of S^{VI} with four O^{2-} ligands. It would then be written in the form $[SO_4]^{2-}$, but generally it is not regarded as necessary to use the square brackets, because sulfate is such a common oxo-anion. The position with regard to some other polyatomic ions such as $[ICl_4]^-$ is not so clear-cut. $(ICl_4)^-$ and ICl_4^- may be equally as acceptable as $[ICl_4]^-$, depending on the precise circumstances of use.

For certain species it may not be obvious how to define a central atom. For example, for chain species such as thiocyanate, the symbols are cited in the order in which the atoms appear in the chain.

Examples

14. –SCN
15. –NCS
16. HOCN
17. HCNO
18. $(O_3POSO_3)^-$

Addition compounds are represented by the formulae of the individual constituent species, with suitable multipliers that define the appropriate molecular ratios, and separated by centre dots.

Examples

19. $3CdSO_4 \cdot 8H_2O$
20. $8H_2S \cdot 46H_2O$
21. $BF_3 \cdot 2H_2O$

These suggestions are advisory and should be used where there are no overriding reasons why they should not. For example, PCl_3O is a correct presentation but, because the $P{=}O$ group persists in a whole family of compounds, in certain contexts the presentation $POCl_3$ may be more useful. There is no objection to this kind of variation.

3.7 GROUPS IN ORGANIC NOMENCLATURE

The concept of group is especially important in organic chemistry. A characteristic (or functional) group is a set of atoms closely linked with chemical reactivity and defined classes of substances. For instance, the hydroxy group, HO– or –OH, is characteristic of the compound classes of alcohols, phenols, and enols. Alcohols are often represented by the general formula R–OH or ROH, in which R– often represents a hydrocarbon group typical of aliphatic and alicyclic substances. The groups R in Table 3.1 are derived from aliphatic hydrocarbons.

A characteristic group occurs in a wide variety of compounds and confers upon them a common kind of reactivity. Table 3.1 shows a selection of common characteristic groups, and organic nomenclature in general is introduced in Section 4.3. Phenols are generally represented by Ar–OH, in which Ar– signifies an aromatic skeleton, typically composed of benzene rings or substituted benzene rings. Enols are molecules in which the –OH group is linked to a carbon atom that is also engaged in a double bond.

Table 3.1 Some important compound classes and characteristic groups

Class	Characteristic Group	General Constitution
Alkanes		C_nH_{2n+2}
Alkenes		$R_2C=CR_2$
Alkynes		$RC\equiv CR$
Alcohols	$-OH$	$R-OH$
Aldehydes	$-\overset{\mid}{C}\overset{H}{\underset{O}{\diagdown}}$	$R-CHO$
Amides	$-\overset{\mid}{C}\overset{NH_2}{\underset{O}{\diagdown}}$	$R-CONH_2$
Amines	$-NH_2, -NHR, -NR_2$	$R-NH_2$ $R-NHR$ $R-NR_2$
Carboxylic acids	$-\overset{\mid}{C}\overset{OH}{\underset{O}{\diagdown}}$	$R-COOH$
Ethers	$-O-$	$R-O-R$
Esters	$-\overset{\mid}{C}\overset{OR}{\underset{O}{\diagdown}}$	$R-COOR$
Halogen compounds	$-F, -Cl, -Br, -I$	$R-F, R-Cl,$ $R-Br, R-I$
Ketones	$>C=O$	$R-CO-R$
Nitriles	$-C\equiv N$	$R-C\equiv N$

Note In this Table, and in common organic usage, R represents an aliphatic group, but in every case in this Table it may be replaced by Ar, which represents an aromatic group, when the compound classification may change, an alcohol then being designated a phenol, for example. In this context, Ar does not represent the element of atomic number 18, argon. In addition, in several items listed in this Table, R may also be hydrogen.

19

Examples

Typical alcohols

1. 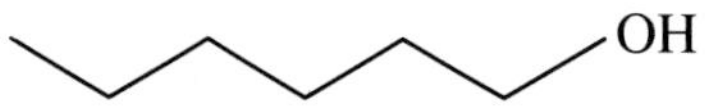   2.

Typical phenols

3. 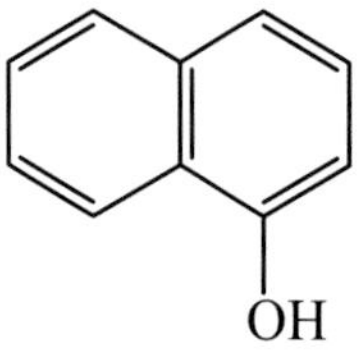   4. 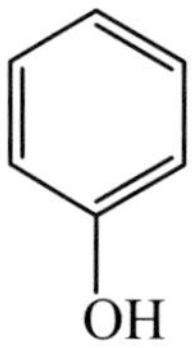   5.

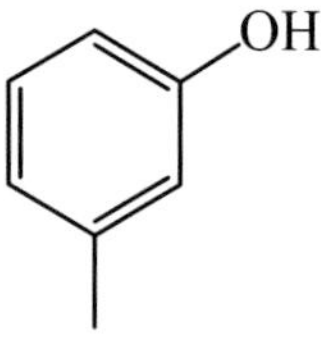

A typical enol

6.

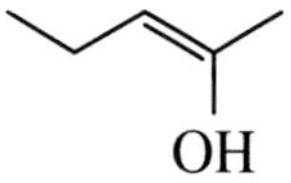

Selecting an appropriate structure upon which to base a name is not always straightforward. Consider the Examples 7 and 8 below. Both structures represent amines. Both represent groups R with seven carbon atoms. Both have an empirical formula $C_7H_{15}NH_2$. However, neither would be named correctly as a heptane. For instance, Example 7 is named as a derivative of pentane whereas Example 8 is named as a derivative of hexane. There is a problem in determining the most senior hydrocarbon chain where several possibilities seem to be present. The choice of parent compound upon which to base a name can become quite complex, and to facilitate the correct choice in acyclic compounds the rules outlined in Table 3.2 should be used. This Table deals with various hydrocarbons, straight-chained and branched, saturated and unsaturated. The selection of parents for further classes of compound is outlined in Table 6.1. This choice will determine the name and also how one writes the structural formula. Some of the steps will only be understood after considering the material presented in Chapter 6.

Examples

7. 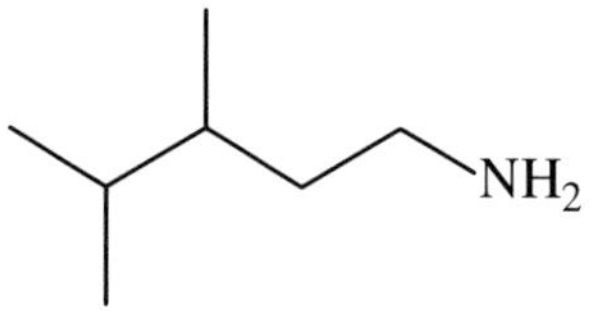   8.

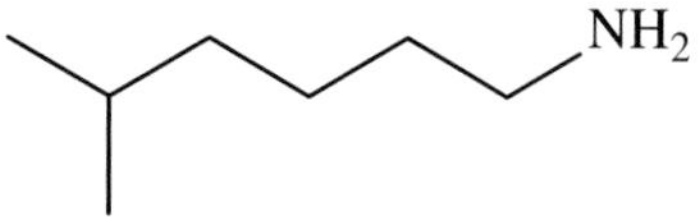

The formulae discussed so far rely on a minimum of structural information. Increasingly, chemists need to convey more than a list of constituents when providing a formula. They need to say something about structure. To do this, simple formulae written on a single line, as in text, need to be modified. How they are modified is determined by what information needs to be conveyed. Sometimes this can be done by a simple modification of a formula in an organic or inorganic ring compound, to show extra bonds not immediately apparent.

Table 3.2 Seniority of chains (the principal chain)

When a choice has to be made of principal chain in an acyclic compound, the following criteria are applied successively, in the order listed, until a single chain is left under consideration. This is then the principal chain.

(a) Select the chain(s) that has (have) the maximum number of substituents corresponding to the principal group, which is the group cited as a suffix.

(b) If this is not definitive, select the chain with maximum length.

If (a) and (b) together are not definitive, the following criteria are applied in order until only one chain remains under consideration.

(c) Chain that has the maximum number of double and triple bonds.

(d) Chain with maximum number of double bonds.

(e) Chain with lowest locants for the principal groups, which are those cited as suffixes.

(f) Chain with the lowest locants for multiple bonds.

(g) Chain with lowest locants for double bonds.

(h) Chain with the maximum number of substituents cited as prefixes.

(i) Chain with lowest locants for all substituents in the principal chain and cited as prefixes.

(j) Chain with the substituents that come first in alphabetical order.

Examples

9. $[Ni\{S\!=\!P(CH_3)_2\}(C_5H_5)]$ 10. $ClCHCH_2CH_2CH_2CH_2CH_2$

Note that these bond indicators do not imply long bonds. Their size and form are dictated solely by the demands of the linear presentation.

For a coordination compound, it is usual to write the formula of a ligand with the donor (or ligating) atom first. The nickel complex in Example 9 has both S and P bonded to the metal, forming a ring. All the carbon atoms of the C_5H_5 are also bonded to the metal. In contrast, the ring structure for chloro-cyclohexane, Example 10, should be immediately obvious.

It may not always be possible to show all the necessary detail in a simple one- or two-dimensional formula. In such cases, attempts must be made to represent three-dimensional structures in two dimensions, as detailed below in Section 3.8.

3.8   THREE-DIMENSIONAL STRUCTURES AND PROJECTIONS

To represent consistently the three-dimensional structure of a molecule or ion in two dimensions it is necessary to work through a fixed series of procedures. The method adopted is to imagine each atom (or group of atoms, if they effectively all occupy the same position, as do the organic characteristic groups listed in Table 3.1) to be placed at the vertex of an appropriate polyhedron. In organic chemistry, this is usually the tetrahedron, with a carbon atom placed at its centre. Examples 1 shows three-dimensional representations of the molecule of methane, CH_4. In (a), the dotted lines represent bonds between carbon and hydrogen, and the full lines show the tetrahedron. In (b) the formalisms of Example 3 of Section 3.4 above are employed.

Examples

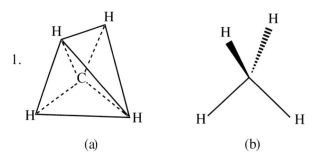

1.

(a)   (b)

This tetrahedral arrangement is quite general. If the four hydrogen atoms were replaced by four chlorine atoms to give CCl_4, a similar tetrahedron would be defined.

For many other organic compounds and for coordination compounds generally, other shapes are taken up. In coordination chemistry these shapes are called coordination polyhedra, a name which was developed historically, though mathematicians might hesitate to call them all polyhedra. A typical coordination complex is shown in Examples 2.

Examples

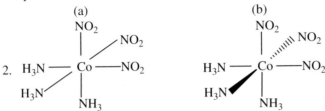

2.

Examples 2(a) and 2(b) show three-dimensional presentations of the coordination complex $[Co(NH_3)_3(NO_2)_3]$ which has a cobalt atom at the central position and six ligand atoms bound at each of six apexes. The use of wedge bonds to emphasise the three dimensions in Example 2(b) is precisely similar to that in Section 3.4, Example 3, cited above. Examples 3(a) and 3(b) below show how the ends of the bonds in (a) may be joined up to form an octahedron (b), in which the dashed lines represent edges which would not be visible if the octahedron were solid.

Examples

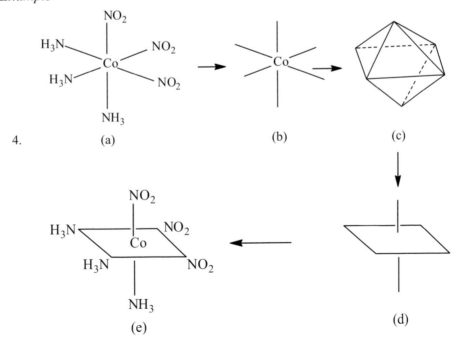

3.

Such polyhedra are often drawn in a highly formalised fashion, which can be confusing to the student. An octahedron is often represented by its apexes, rather than by its faces. The structure (a) in Examples 4 gives rise formally to the octahedron in (c). The lines defining the central plane in (d) are edges of the parent octahedron, the lines perpendicular to this plane connect the central atom to the two remaining vertices. The resultant presentation is shown in (e).

Example

4.

However, Example 4(e), which is equivalent to Example 4(a), is not intended to indicate bonds between, for example NH_3 and NO_2, merely the shape of the compound. The representation is perhaps an unfortunate hybrid of a three-dimensional representation and a line formula in which only selected bonds are shown. Care needs to be exercised when using this format and it is not recommended for elementary purposes unless it is required to illustrate a particular point.

Normally a two-electron bond is represented in these formulae by a line. When electron pairs are not conveniently localised between specific atom pairs, as in aromatic compounds, it is not entirely accurate to represent bonds in this way. For example, benzene as drawn in Example 5(a) is sometimes represented in ways such as those shown in Examples 5(b) and 5(c).

Examples

5.

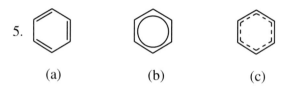

      (a)   (b)   (c)

In coordination compounds, similar representations for delocalisation are often used, as in the depiction of ferrocene (Example 6). Note also the formalism adopted to portray the cyclopentadienyl rings.

Example

6.

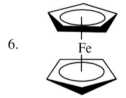

Another way to represent a three-dimensional object in two dimensions is to use some kind of projection. These are used particularly in organic chemistry. In a Fischer projection, generally used to represent carbohydrates and related molecules, the atoms or groups of atoms attached to a tetrahedral centre are projected onto the plane of the paper in such an orientation that atoms or groups of atoms appearing above or below the central atom lie behind the plane of the paper, and those appearing on either side of the central atom lie in front. It is mandatory initially to 'set up' the molecule in an appropriate configuration. If there is a main carbon chain, it is always aligned vertically.

Examples

7.

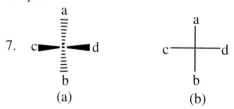

      (a)   (b)

A three-dimensional structure is shown in Example 7(a), the hashed wedges indicating bonds directed to groups behind the plane of the paper, and the solid wedges indicating bonds pointing to groups in front of it. The Fischer projection is shown in Example 7(b). Note that some authorities prefer to use a thickened line rather than the wedge bond to represent a bond projecting towards the reader, but this usage is not recommended by IUPAC. Organic practice is never to indicate a carbon atom in a projection by an atomic symbol. Fischer projections were originally designed to represent the complicated stereo-chemistry of carbohydrates, and it is recommended that their use be restricted to representing these and similar species, such as amino acids.

Another type of projection, a Newman projection, is often employed. This is obtained by viewing the molecule under consideration along a bond, as in Examples 8, which shows how the Newman projection 8(c) is developed.

Examples

8.

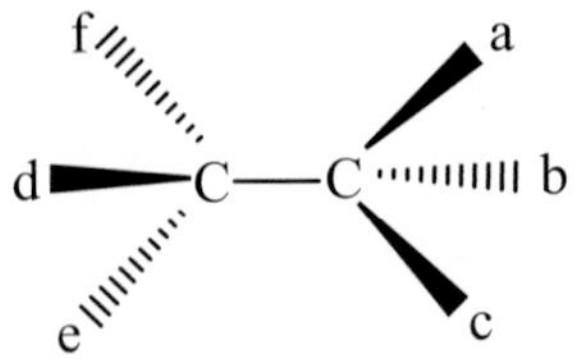

(a) 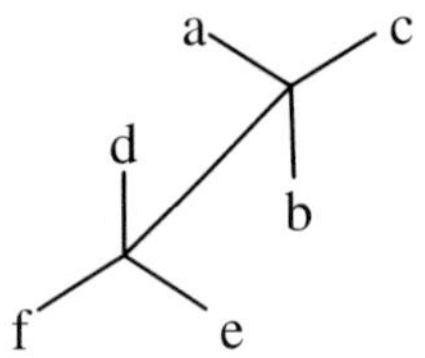   (b)

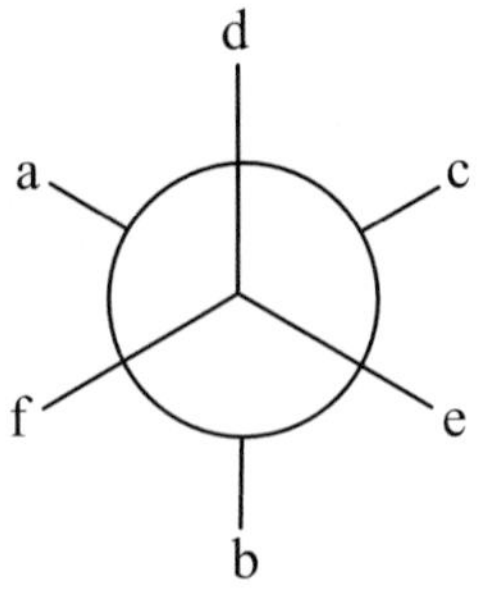

(c)

Structure 8(a) is a three-dimensional presentation of an ethane derivative, $C_2abcdef$, the full line indicating a bond in the plane of the paper in the usual way, and the wedges showing the orientation of the bonds to the substituents. Structure 8(b) is a perspective view of this molecule. Finally, structure 8(c) is the Newman projection. The point at the centre of the circle represents the carbon atom nearest the reader, and the second carbon atom of the ethane backbone is represented by the circle. Lines from outside the circle represent bonds from other atoms to the further carbon atom; such bonds to the carbon atom nearer the reader meet at the centre. When such bonds would be coincident in a projection, they are drawn at a small angle to each other. If the carbon-carbon bond allows rotation of the groups attached to the carbon atoms of the carbon-carbon bond, then the possibility of an infinite number of conformations arises, but not all are of the same energy. Two specific ones corresponding to different rotations about the C–C bond are termed eclipsed and staggered. They are shown in the perspective drawings represented in Examples 9(a) and 9(b). Newman projections can distinguish between these, as shown in representations in Examples 9(c) and 9(d). The former, Example 9(a), is eclipsed, bringing the substituent atoms in closer proximity than in the latter conformer 9(b), and the latter is of lower energy.

Examples

9.

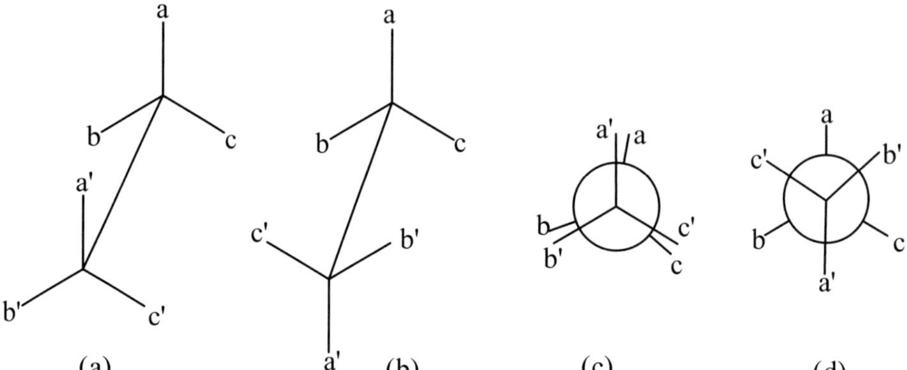

(a)   (b)   (c)   (d)

Other conformations encountered in the literature are categorised in terms of the Newman projections, as shown in Examples 9. Note that the terms *syn* and *anti* are no longer recommended in this context. The chlorine atoms in Examples 10 may be described as synperiplanar, synclinal, anticlinal or antiperiplanar to each other.

Examples

10.

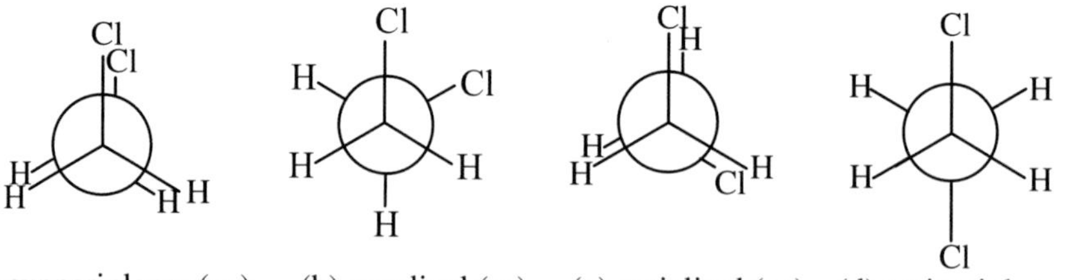

(a) synperiplanar (sp)   (b) synclinal (sc)   (c) anticlinal (ac)   (d) antiperiplanar (ap)
                          or gauche

Newman projections may be useful for portraying molecules of complex alkanes and their derivatives, but they are not applied generally. **Haworth projections** are used to represent the structures of monosaccharides (see Chapter 11). **Zig-zag projections** for representing the tacticity of polymers may be used where necessary, but will not be commonly met with by elementary students and are not further dealt with here, though more information can be found in: Basic terminology of stereochemistry, IUPAC Recommendations 1996, *Pure and Applied Chemistry*, 1996, **68**(12), 2193–2222.

In inorganic compounds, configurational arrangements other than octa-hedral and tetrahedral are commonly observed. A selection of polyhedra encountered in organic and inorganic chemistry is shown in Table P4. A further selection of polyhedra found principally in coordination chemistry is shown in Table P5. The latter also includes the **polyhedral symbols** that denote each **polyhedron** and also the corresponding **coordination number**. These terms are discussed more fully in Chapter 7.

Note that in Table P4 a hybrid stereo view of a structure is often used, in which some lines represent bonds and others edges of the polyhedron that defines the shape, just as discussed above for the octahedron. The same caution as that exercised for the octahedron should be used with the representations shown in Table P4. The central atom is indicated here by the letter M, and the attached groups by letters a, b, c, *etc.* For a given formula, for example for a general formula [Mabcde], more than one shape or configuration may be possible. Tables P4 and P5 should be considered together.

3.9 ISOMERS AND STEREOISOMERS

3.9.1 Introduction

Isomerism describes the relationship between molecular entities having the same molecular formula, but differing in structure and/or connectivity between con-stituent atoms. For example, the molecular formula C_7H_{16} corresponds to many different alkanes that differ from each other in their bonding. Two are shown in Example 1(a). In the same manner, various structural formulae can be envisaged for the molecular formula C_3H_6O, Example 1(b), one being a ketone and another an alcohol.

Examples

1.

(a) and

(b) CH_3COCH_3 and $CH_2=CHCH_2OH$

Stereoisomers are isomers having the same molecular formula and the same bonding connections, but different spatial arrangements of atoms. Classes of stereoisomer include *cis-trans* isomers, conformational isomers, and enantiomers.

3.9.2 *cis-trans* Isomers and stereodescriptors

The stereodescriptors *cis* (*i.e.*, on the same side) and *trans* (*i.e.*, on opposite sides) indicate the spatial distribution of atoms or groups with respect to a plane defined by the molecular structure, often in relation to a double bond which does not allow free rotation about itself. In Example 2(a), the substituents a are *trans* to each other, whereas in 2(b) they are *cis*.

Examples

2.

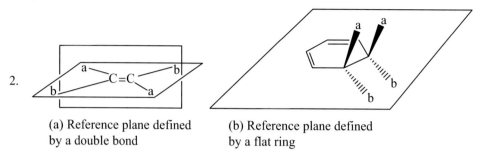

(a) Reference plane defined by a double bond

(b) Reference plane defined by a flat ring

The stereodescriptors *cis* and *trans* are acceptable for simple organic structures, as shown in Examples 3.

Examples

3.

(a) *cis*-5,6-dimethylcyclohexa-1,3-diene   (b) *trans*-5,6-dimethylcyclohexa-1,3-diene

The italic letters *Z* and *E* are sometimes used as synonyms for *cis* and *trans*, as illustrated in Examples 4(a) and 4(b). In more complex cases the *cis* and *trans* stereodescriptors may be ambiguous and it is now generally recommended that they be replaced by the descriptors *Z* and *E*, though not for ring systems. The meanings may be remembered by reference to the German words 'zusammen' (together) and 'entgegen' (opposite).

Examples

4.

(a) *cis*-but-2-ene **or** (*Z*)-but-2-ene   (b) *trans*-but-2-ene **or** (*E*)-but-2-ene

Designating a plane using *Z* and *E* for inorganic structures such as the square plane and the octahedron is not encouraged. The descriptors *cis* and *trans* have also been used to describe spacial distribution in octahedral and square planar structures such as coordination complexes. For square planar complexes of the general formulae [Ma$_2$b$_2$] and [Ma$_2$bc] this can work quite well, as shown in Examples 5.

Examples

5.

(a) *cis*-diamminedichloridoplatinum(II)   (b) *trans*-diamminedichloridoplatinum(II)

However, these descriptors are not adequate to distinguish all possibilities even in square planar complexes [Mabcd], and in octahedral coordination a simple descriptor cannot account for all possible arrangements because all the ligands are simultaneously *cis* and *trans* to each other. The system that is currently recommended for coordination compounds is described in more detail in the *Nomenclature of Inorganic Chemistry, IUPAC Recommendations 2005,* Chapter 9, and employs the **configuration index** or CI. This can be applied to many regular polyhedra.

The configuration index is derived by applying the Cahn-Ingold-Prelog or CIP rules (reference 2) to all the donor (or ligating) atoms in order to assign to each a priority number, generally based upon atomic number, the larger the atomic number, the higher the priority. These priorities are used in both organic and inorganic nomenclatures. The first CIP rule states that a substituent (or ligating atom in this kind of application) of higher atomic number is senior to a substituent (or ligating atom) of lower. For example, in the square planar complexes $[PtCl_2(NH_3)_2]$ in Examples 5 above, the ligand atom Cl has a priority 1 and the ligand atom N has the priority 2. The configuration index is selected as the priority of the ligating atom trans to the ligating atom of highest priority. This gives the configuration indices shown in Examples 6.

Examples

6.

(a) *cis*-diamminedichloridoplatinum(II)
N, priority 2, opposite Cl, priority 1
CI = 2, combined with polyhedral symbol
gives *SP*-4-2
Revised name
(*SP*-4-2)-diamminedichloridoplatinum(II)

(b) *trans*-diamminedichloridoplatinum(II)
Cl, priority 1, opposite Cl, priority 1
CI = 1, combined with polyhedral symbol
gives *SP*-4-1
Revised name
(*SP*-4-1)-diamminedichloridoplatinum(II)

In Examples 6, the CI displays no particular advantage over the descriptors *cis* and *trans*, but in more complex systems the CI is necessary to define the configuration exactly. The student should be able to recognise what the device is, but only the more advanced will need to learn how to derive and apply it.

3.9.3   **Conformational isomers or conformers**

The conformation of a molecule is the spatial arrangement of the atoms. A conformer is one of a set of molecular arrangements of similar atom content and connectivity, and differing from one another in their conformations, each of which is considered to correspond to a potential-energy minimum. Different stereoisomers that can be interconverted by rotation about a single bond are called conformers. The interconversion of conformers by rotation about a single bond involves crossing an energy barrier between different potential-energy minima, as described in Section 3.8, Examples 9 and 10.

Examples

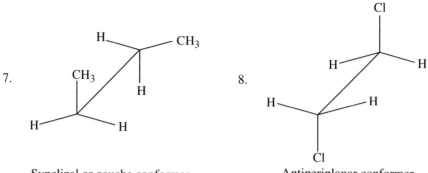

7.   8.

Synclinal or gauche conformer   Antiperiplanar conformer

The concept of conformational analysis has led to a better understanding of the spatial arrangements of cyclic alkanes, and of the chemical reactivity of functionalised derivatives. A specific terminology is used.

Examples

9. Envelope conformation   10. Chair conformation   11. Boat conformation   12. Twist conformation

In a cyclohexane or similar molecule, extra-skeletal bonds are described as equatorial (e) or axial (a) (Example 13). Example 13(b) shows the same molecular shape, but emphasises the three-dimensional structure of the ring.

Example

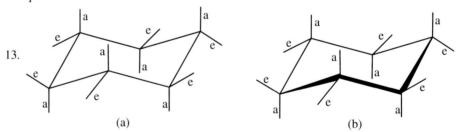

13.

(a)   (b)

3.9.4   Enantiomers

The property of an object of which an image is not identical with (or superimposable upon) its mirror image is termed **chirality**. For example, the human right hand has the same shape and internal structure as the human left hand, but they are different non-superimposable objects. They are mirror images of each other. When two such molecules exist in chemistry, they are termed **enantiomers**. Enantiomers have identical physical properties (except for their interaction with polarised light) and chemical reactivity (except for reaction with other chiral species). Consequently, any biological activities that involve stereospecificity may also be different. The directions of the specific rotations of polarised light of the two enantiomers are of equal magnitude but in opposite directions. A **chiral molecule** is not superimposable on its mirror image, whereas an **achiral molecule** is superimposable. Chirality is due to the presence in a molecule of a centre, axis, or plane of chirality. Only chirality centres will be dealt with here.

A **chirality centre** is an atom which is bonded to a set of other atoms or groups (these may also be termed ligands, especially in inorganic chemistry) in a spatial arrangement that is not superimposable on its mirror image. For example, a carbon compound Cabcd, a phosphorus compound Pabc, and an ammonium ion $(Nabcd)^+$ are chiral when the entities a, b, c, and d are all different. Example 14 shows a typical **enantiomeric pair** (see Section 3.9.4).

Example

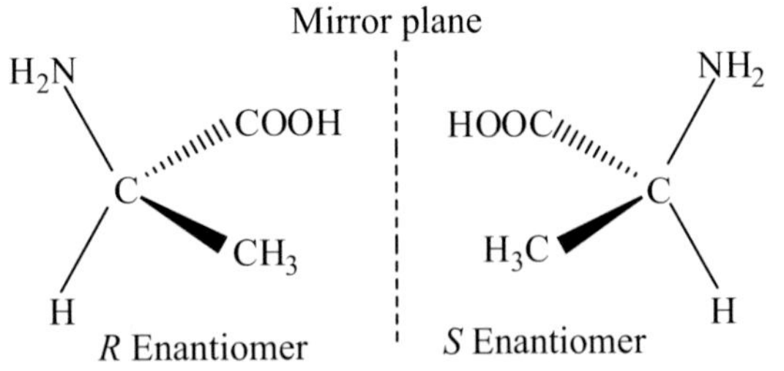

R and S were assigned using CIP priorities

The symbols *R* and *S* are termed **stereodescriptors**, and are used to distinguish between the enantiomers in structures which can be regarded as tetrahedral. These are assigned using the Cahn-Ingold-Prelog (CIP) priorities conferred upon the different substituents a, b, c, and d (reference 2 and Section 3.9.2). As described above, the first CIP rule states that a substituent of higher atomic number is senior to a substituent of lower. The assignment of descriptors *R* and *S* in Example 14 was made as follows.

Consider a molecule Cabcd, where the alphabetical order represents the relative seniorities of the groups. In Examples 14, a, b, c, and d were assigned by considering the atomic numbers of the individual atoms bound to the central carbon atom, which are C in CH_3, C in COOH, H, and N in NH_2. In decreasing order: N = a, C = both b and c, and H = d. There are further CIP rules for cases such as this, and these dictate that COOH = b, and CH_3 = c.

Example

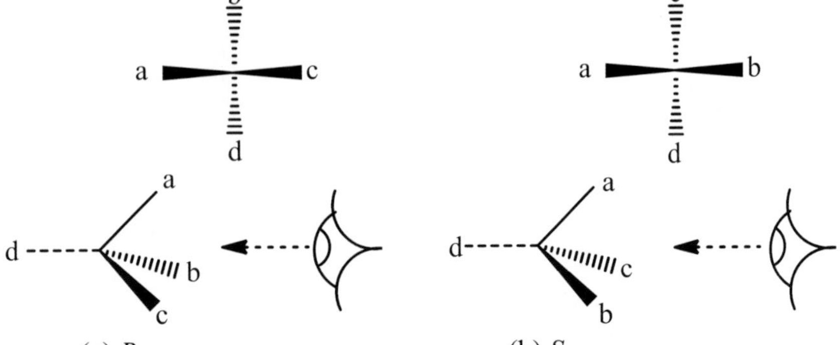

The tetrahedral molecule in Example 14 is first set up as shown in Examples 15. The molecule is then formally viewed along the bond connecting the carbon atom in the chirality centre to d, the group of lowest CIP priority, but from the side remote from d, as shown in Examples 15(a) and 15(b) above. If the remaining sequence a, b, c is clockwise, then the enantiomer considered is assigned the descriptor *R* (from the Latin *rectus*, right). If it is anti-clockwise, it is assigned the descriptor *S* (from the Latin *sinister*, left). Several further rules are required to treat every possible eventuality, but it is not necessary to consider them here.

When a molecule contains two tetrahedral chirality centres, as in a hypothetical molecule $C_2abcdef$, three or four steroisomers are possible, depending on whether or not a plane of symmetry or inversion centre of the whole molecule is present in one of them. In a set of four isomers, there are two pairs of enantiomers. Stereoisomers that are not enantiomers of each other are called **diastereoisomers**. This is illustrated in Examples 16. Compounds I and II are enantiomers and compounds III and IV are also enantiomers, but compounds I and III, for instance, are diastereoisomers.

Example

16.

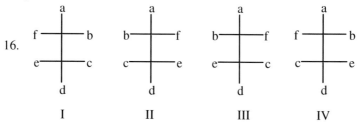

I   II   III   IV

A plane of symmetry or an inversion centre removes the chirality, making a molecule **achiral**, and this reduces the number of isomers. Examples 17 show a pair of enantiomers, but Example 17(b) has a plane of symmetry and is termed a ***meso*** compound. Its specific rotation of polarised light is $0°$ so it does not rotate the plane of polarised light, unlike the two individual enantiomers in Example 17(a). A mixture of equimolar proportions of enantiomeric molecules, such as in Example 17(a), cancels out the rotations of the polarised light due to the separate enantiomers, and also has a specific rotation of polarised light of $0°$. Such a mixture is termed a **racemate**. The constituents of a racemate may be difficult to separate.

Examples

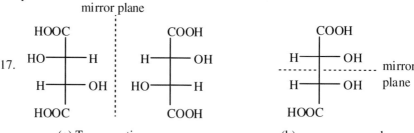

(a) Two enantiomers   (b) *meso* compound

Examples 18–23 below demonstrate further the uses in organic nomenclature of the stereodescriptors described above. The detailed methods of deriving the names and locants will be described in Chapter 6. Note the alternative presentations of the alkyl groups. Either is acceptable as long as the structural diagram depicted is unambiguous.

Examples

18.

(*E*)-pent-2-ene

19.

(*Z*)-pent-2-enoic acid

20.

21.

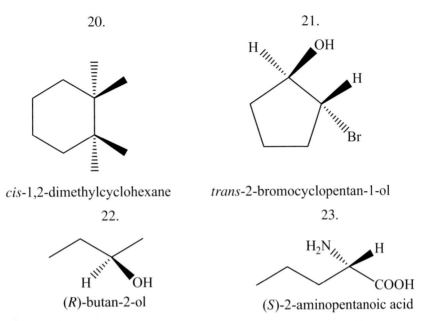

cis-1,2-dimethylcyclohexane

trans-2-bromocyclopentan-1-ol

22.

23.

(*R*)-butan-2-ol

(*S*)-2-aminopentanoic acid

Clearly, for inorganic and organometallic enantiomeric complexes with geometries that cannot be regarded as tetrahedral, the descriptors *R* and *S* cannot be applied. For this reason, a modified but analogous approach has been developed, in which the descriptors *C* (for clockwise) and *A* (for anticlockwise) are employed. Two polyhedral examples are considered here simply to exemplify the methodology. Beginning students should not be expected to derive or use these symbols, but they should be able to recognise what they represent. Full details are contained in the *Nomenclature of Inorganic Chemistry, IUPAC Recommendations 2005*, Chapter 9.

The procedure is similar to that described above for tetrahedral molecules. The molecule is set up so that it is viewed along the principal axis with the donor (or ligating) atom of higher priority oriented towards the viewer. For a trigonal bipyramidal molecule this is shown in Example 24.

Example

24.　　　Principal axis

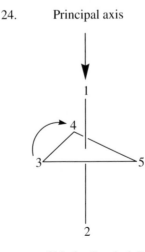

Chiralty Symbol *C*

33

In this case, the priority sequence in the plane of the approximately equilateral triangle is $3 \rightarrow 4 \rightarrow 5$ and clockwise. Therefore the configuration is *C*. Were 4 and 5 to be interchanged, the configuration would be *A*.

Examples 25 illustrate the more complicated octahedral case. There are four ligands in the square plane and as they are all different they can take up six possible arrangements, three of which are shown below.

Examples

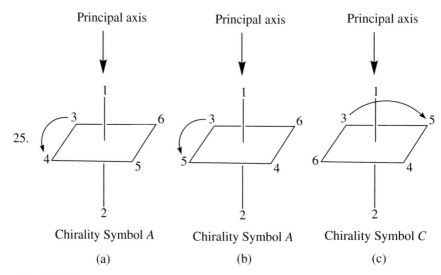

Chirality Symbol *A*	Chirality Symbol *A*	Chirality Symbol *C*
(a)	(b)	(c)

Considering Example 25(a), since the sequence $3 \rightarrow 4$ is lower than $3 \rightarrow 6$, then this is anticlockwise and the chirality symbol is *A*. If the donors (ligating atoms) of priority 6 and 4 were interchanged, then the chirality would have been designated *C*. Similarly, in Example 25(b), $3 \rightarrow 5$ takes precedence over $3 \rightarrow 6$, so this is equally *A*. Finally, in Example 25(c), $3 \rightarrow 5$ takes precedence over $3 \rightarrow 6$, so this is designated *C*. More complex cases are dealt with in the *Nomenclature of Inorganic Chemistry, IUPAC Recommendations 2005*, but only advanced students and researchers are likely to need more detail than supplied here.

REFERENCES

1. Graphical representation of stereochemical configuration, Recommendations, *Pure and Applied Chemistry*, 2006, **78**(10), 1897–1970; Graphical representation standards for chemical structure diagrams, Recommendations *Pure and Applied Chemistry*, 2008, **80**(2), 277–410.
2. R.S. Cahn, C. Ingold and V. Prelog, *Angew. Chem., Int. Ed. Engl.*, 1966, **5**, 385–415 and 511; V. Prelog and G. Helmchen, *Angew. Chem., Int. Ed. Engl.* 1982, **21**, 567–583. *A Guide to the IUPAC Nomenclature of Organic Compounds, Recommendations 1993*, Blackwell Science, Oxford, pp. 152–154; *Nomenclature of Organic Chemistry*, Pergamon Press, Oxford, 1979, pp. 486–490.

4   Types of Nomenclature

INTRODUCTION

Specialists in nomenclature recognise two different categories of nomenclature, *trivial* and *systematic* (see Chapter 2). In nomenclature, 'trivial' is not a dismissive or pejorative term. It indicates names that are arbitrary and not derived by systematic methods. Trivial names include the names of the elements, such as sodium and hydrogen, as well as everyday and laboratory names such as caffeine, diphos and LithAl. Trivial nomenclature contrasts with systematic nomenclature, which is an assembly of naming rules, though these rules may be themselves arbitrary. It has already been noted that the names of the elements, trivial in themselves, are the basis of systematic nomenclature.

One function of nomenclature specialists is to codify such rules so that everyone can use them to identify pure substances, rather as many people use an alphabet to represent words. There may be more than one way to name a compound or species, and no one way may be superior to all the others. In the past different systems have been developed to suit the particular needs of individual groups of chemists. Names also vary in complexity, depending upon how much information needs to be conveyed. For example, a compositional name conveys less information than a structural or constitutional representation of a molecule, because it contains no information about the arrangement of atoms in space.

From the beginning of chemistry, chemists and alchemists developed names for materials that were trivial. For example, many so-called vitriols (blue, green, oil of) were understood to share a common factor, which today we recognise as the sulfate ion. Now that we understand much more about the way in which atoms combine, we can construct names which can give information about stoichiometry and structure. Nevertheless, many usages that have their roots in the distant past are still embedded in nomenclature. In addition, there are several widely used systems of nomenclature, and these tend to reflect the kinds of chemistry for which they were developed.

IUPAC concerns itself principally with compositional, substitutive, and additive nomenclatures. Various other nomenclatures may be found in the literature, such as enzyme nomenclatures, pharmaceutical nomenclatures, *etc*. A short discussion of some of these is presented in Chapter 12.

4.2   COMPOSITIONAL NOMENCLATURE

This is a system based upon stoichiometry. It is not restricted to binary (two-element) compounds, but the nomenclature is often binary in structure, as discussed below.

Principles of Chemical Nomenclature: A Guide to IUPAC Recommendations
Edited by G. J. Leigh
© International Union of Pure and Applied Chemistry 2011
Published by the Royal Society of Chemistry, www.rsc.org

CHAPTER 4

4.3   SUBSTITUTIVE NOMENCLATURE

This is the principal nomenclature used in organic chemistry, as described in the *Guide to IUPAC Nomenclature of Organic Compounds, Recommendations 1993* and the *Nomenclature of Organic Chemistry, 1979.* Updates of some of this material are available on-line at http://www.iupac.org.

This system of nomenclature is based upon the name of a formal parent molecule, normally a so-called **parent hydride**, which is then substituted in a formal sense, that is, one or more hydrogen atoms are replaced by other moieties, or the parent hydride may be formally charged by adding or removing charged entities. Such changes are denoted by modifying the parent name with **suffixes** and/or **prefixes** and/or **infixes** (together termed **affixes**). Examples of such suffixes and prefixes, which are sometimes also used in other nomenclature systems, are 'ium' and 'hexa'.

Substitutive nomenclature follows the pattern established by the original development of organic chemistry. Although it is principally an organic nomenclature, it is now of more general use and is often applied to the names of parent hydrides of the elements of Groups 14, 15, 16, and 17 and their derivatives.

4.4   ADDITIVE (OR COORDINATION) NOMENCLATURE

Additive nomenclature was developed originally to name **coordination compounds**, although, when appropriate, it is now also used in other circumstances. It is sometimes called coordination nomenclature. For a more detailed discussion, see the *Nomenclature of Inorganic Chemistry, IUPAC Recommendations 2005*. A compound to be named is considered to consist of a **central atom** together with its **ligands**, and the name is developed by assembling the individual names of the constituents. In this way, the system parallels the ideas developed shortly after the beginning of coordination chemistry, when it was inferred that each ligand donated an electron pair to a metal acceptor atom to form a bond. This system was later also applied to name oxoacids and related anions. Coordination names for oxoanions are cited in the Examples throughout this book and they are presented in detail in Chapter 7.

4.5   POLYMER NOMENCLATURE

Polymers are bulk materials made up of **macromolecules**, which generally consist of repeating units of the same or of a limited number of kinds, bound together in branched or unbranched chains to form macromolecules of potentially infinite length. They differ from the molecules discussed so far in that the latter have been of limited, often strictly defined, size, and their nomenclature is based upon the systems described above. The molecules of a polymer may not have a uniform structure, and a given polymer may consist of a mixture of macromolecules of different lengths (**degrees of polymerisation**). Each macromolecule in a polymer may also have a different structural arrangement from the others in the mixture. Such peculiarities of polymers led to the recognition of two princial nomenclature systems, termed **structure-based nomenclature** and **source-based nomenclature**.

As the name implies, structure-based nomenclature attempts to describe the structure of the molecules in a polymeric material. Because the individual macromolecules in a polymeric sample may differ from each other in a number of ways (irregularities in monomeric unit sequences within chains, branching irregularities, monomeric unit orientation, end-group structure, stereo irregularities, *etc.*), a complete description of a polymer structure is very difficult and structure representations are generally idealised. Precise definition of a polymeric substance using a structure-based name is therefore often impossible.

In contrast, deriving a name for a polymer by describing the monomer from which it was derived is generally possible. This is the basis for source-based nomenclature. Many names for polymers such as poly(vinyl chloride) are source-based. Even this may be insufficient to derive an unequivocal name for a polymer. In such circumstances a **generic source-based** nomenclature may be of use.

Source- or structure-based names can be formulated for most polymers using these nomenclature systems. For more information, see Chapter 9.

4.6 BIOCHEMICAL NOMENCLATURE

Biochemistry is a multidisciplinary subject, and consequently its nomenclature is governed by rules which differ somewhat from those laid down by IUPAC for organic and inorganic chemistry. In the case of biochemical nomenclature, these rules are decided jointly by IUPAC and the International Union of Biochemistry and Molecular Biology (IUBMB). Biochemical subjects are more and more frequently investigated by chemists, so it was thought appropriate to include material on biochemical nomenclature in this book, in order to describe the names of the kinds of compound often met with in chemical literature. This will be found in Chapter 11.

4.7 OTHER SYSTEMS

Several systems of nomenclature are more restricted in their application and sometimes make use of trivial, traditional terms. They include **acid nomenclature** (generally inorganic, and often used for oxoacids and derivatives), **replacement nomenclature** (mainly organic, to denote replacement of skeletal atoms in a parent hydride rather than replacement of hydrogen atoms, as in the use of the infixes 'oxa' and 'aza' to indicate replacement of a skeletal carbon atom by oxygen and nitrogen atoms, respectively) and **functional class nomenclature** (this is also primarily organic and involves the use of class names such as alcohol, acid, and ether), **inorganic rings and chains nomenclature**, and **subtractive nomenclature** (such as organic **deoxy** and inorganic **debor** nomenclatures) These will be referred to briefly where appropriate.

More than one naming system may be applied to the same compound. For example, $SiCl_4$ can be named silicon tetrachloride (binary name), tetrachlorosilane (substitutive name, chlorine atoms replacing hydrogen atoms) and even tetrachloridosilicon (additive name). No one name is 'better' or 'more correct' than any other, though IUPAC is currently considering whether one name of a selection available for any given compound should be preferred, primarily in order to simplify identification of compounds for commercial and legal purposes. When such names are finally selected, they will be referred to as **Preferred IUPAC Names** or **PINs**.

5 Compositional and Binary-Type Nomenclatures

5.1 COMPOSITIONAL AND BINARY NOMENCLATURES

Although it is possible to develop a name based simply on an empirical formula (see Chapter 3), this is rarely sufficient unequivocally to describe a compound. In the simplest cases, the species to be named contains only one element, and the name is formed by adding the appropriate multiplicative (or numerical) prefix to the element name. Multiplicative prefixes used here and in other types of nomenclature are listed in Table P6.

Examples
    1. S_8 octasulfur
    2. O_2 dioxygen
    3. O_3 trioxygen (trivial name ozone)

These names are examples of **compositional nomenclature** and are satisfactory to identify compounds such as these, but as soon as more complicated types of compound are considered their limitation becomes clear.

Consider the compound of empirical formula CaO_4S. A reasonable proposal for a compositional name might be calcium tetraoxygen sulfur, but this adds nothing to the bare formula. However, this compound is more immediately recognised by the formula $CaSO_4$, and the name calcium sulfate. This name is in two parts, and is thus classified as a binary name. **Binary nomenclature** is used widely in inorganic chemistry, though it is also employed for organic systems where appropriate.

5.2 BASIS OF THE BINARY SYSTEM OF NOMENCLATURE

When a particularly simple name is required, for example on the label of a bottle, or when the user has little hard structural information to be conveyed in the name, then binary nomenclature can provide adequate names. In particular, using the assumed or established division of constituents into positive and negative parts already discussed in Chapter 3 (see also Table P3) in establishing formulae, the constituents are formally divided into two classes, hence the term 'binary nomenclature'. These classes are termed **electropositive** and **electronegative**.

However, there is no general scale of electropositivity, although electronegativity has been widely discussed in the literature. The constituents of a

Principles of Chemical Nomenclature: A Guide to IUPAC Recommendations
Edited by G. J. Leigh
© International Union of Pure and Applied Chemistry 2011
Published by the Royal Society of Chemistry, www.rsc.org

compound are best considered as being more or less electronegative. As discussed in Chapter 3, even this is not to be interpreted too rigidly, and in nomenclature we are forced to use various arbitrary devices to define relative electronegativities, such as that displayed in Table P3. Though the terms electronegative and electropositive continue to be used in this context, this is because they are sanctioned by long use. In no circumstances should numerical values be applied to such terms when used in nomenclature.

5.3   NAME DERIVATION

A binary name is derived by combining the names of the electropositive constituent(s) with those of the electronegative constituent(s), suitably modified by any necessary multiplicative prefixes. The electropositive constituent names are cited first, and are separated from the electronegative constituent names by a space. Multiplicative prefixes may not be necessary if oxidation states are explicit or are clearly understood. However, oxidation state information should no longer ever be conveyed by using the suffixes 'ous' and 'ic', because they are confusing and they have not been applied consistently. For example ferrous and ferric refer to oxidation states II and III, respectively, whereas cuprous and cupric refer to oxidation states I and II. Elements that exhibit more than two oxidation states pose further problems. However, though the names of acids, such as sulfurous and nitrous and sulfuric and nitric, present the same problem, these are not considered unacceptable because such names have no simple alternatives and are needed in organic nomenclature. Additive names (see Chapter 7) are also available for these compounds, as in Section 5.4, Examples 8 and 9, and in Table 5.1.

Examples
   1. $NaCl$ sodium chloride
   2. Ca_3P_2 calcium phosphide
   3. Fe_3O_4 triiron tetraoxide

The name of an electropositive constituent is simply the unmodified element name (Examples 1, 2, and 7), the name of a polyatomic cation (Example 4), or the accepted name for a group of atoms which retains its integrity in a variety of chemical species (Examples 5 and 8), as appropriate.

Examples
   4. NH_4Cl ammonium chloride
   5. $[UO_2]Cl_2$ dioxidouranium dichloride
   6. $O_2[PtF_6]$ dioxygen hexafluoridoplatinate
   7. OF_2 oxygen difluoride
   8. $NOHSO_4$ oxidonitrogen hydrogensulfate

If there is more than one electropositive constituent, the names should be spaced and cited in alphabetical order of the initial letters, or of the second letters if the initial letters are the same, and so on. Multiplicative prefixes are ignored for the purposes of ordering both electropositive and electronegative constituents, unless they are an integral part of the name of the entity being qualified (as they are in Examples 5 and 6 above).

If hydrogen is present as an isolated positive ion, it should be cited last amongst the electropositive constituents, but this is rare and it is usually found in association with an anion, as in Example 8 above or Example 10 below. Its name precedes the names of its electronegative partner, without a space. In languages other than English, different ordering may apply. In the Examples below the letters defining the order are in bold face for clarity. The use of bold face here is to demonstrate the principle and should not be extended to normal practice.

Examples

9. $KMgCl_3$		**m**agnesium **p**otassium chloride
10. $NaNH_4HPO_4$		**a**mmonium **s**odium hydrogenphosphate
11. $Cs_3Fe(C_2O_4)_3$		tri**c**aesium **i**ron trioxalate
12. $AlK(SO_4)_2 \cdot 12H_2O$		**a**luminium **p**otassium bis(sulfate)—water (1/12)

Example 12 shows how the formula of a compound considered as an addition compound is converted to a name. The molecular proportion is shown as the appropriate ratio (here 1/12) in parentheses after the names, which are themselves separated by a long dash.

The names of monoatomic electronegative constituents are derived by modifying the names of the corresponding elements. The common termination, of the element name, often but not always 'ium', is replaced by the anion suffix 'ide'. This treatment differs from that of the names of electropositive constituents and has no obvious logical basis, but is retained for historical reasons. In certain cases the modification is accompanied by an abbreviation, and there are a few anion names that are based upon Latin or other roots, although the element names are English. Such names are shown in Table P7. The names of polyatomic, homoatomic species are designated using appropriate multiplicative prefixes. Some trivial anion names are still allowed (Examples 18–22).

Examples

13. Cl^-	chloride
14. S^{2-}	sulfide
15. S_2^{2-}	disulfide
16. Sn_9^{4-}	nonastannide
17. I_3^-	triiodide
18. O_2^-	superoxide
19. O_2^{2-}	peroxide
20. O_3^-	ozonide
21. N_3^-	azide
22. C_2^{2-}	acetylide

Note that upon forming a full name, as for Na_4Sn_9, tetrasodium (nonastannide) and $Tl(I_3)$, thallium (triiodide), enclosing marks may be necessary to avoid ambiguity. Comparison of TlI_3, thallium tris(iodide) with $Tl(I_3)$, thallium (triiodide) illustrates the need for this device, the former compound being composed of thallium(III) cations and iodide anions, I^-, whereas the latter is composed of thallium(I) cations and triiodide anions, I_3^-. In some circumstances, such as when the name of the electronegative species is cited alone, for example, when discussing an anion, it may be useful to indicate the charge, as in Examples 15, disulfide(2−), 16, nonastannide(4−), and 17, triiodide(1−). This is discussed further below. If there is more than one electronegative constituent the names are ordered alphabetically, as with the names of the electropositive constituents, and similarly ignoring multiplicative prefixes which are not part of the name.

Examples

   23. KCl   potassium chloride
   24. $BBrF_2$ boron **b**romide di**f**luoride
   25. PCl_3O phosphorus tri**c**hloride **o**xide
   26. $Na_2F(HCO_3)$ disodium **f**luoride **h**ydrogencarbonate

In Example 26, 'sodium' would be just as acceptable as 'disodium' because in most circumstances the necessity for the presence of two sodium cations would be assumed to be obvious. The name 'hydrogencarbonate' contains no space, though organic nomenclature rules require that similar names for organic acids be written with a space, *e.g.*, hydrogen tartrate.

5.4   NAMES OF POLYATOMIC ANIONS

The names of polyatomic negative assemblages containing more than one kind of atom are derived in various ways. Heteroatomic electronegative constituents generally take the anion ending 'ate', which is characteristic of the names of anions of oxoacids, such as sulfate, phosphate and nitrate, and of complex anions such as hexafluoridoplatinate. However, many more of such anions can be regarded as coordination compounds, and their systematic names may then be assembled using the rules of additive nomenclature (Chapter 7). Some traditional trivial names are still allowed, even when they do not carry the 'ate' endings (Examples 1–9).

Examples

   1. CN^- cyanide
   2. NH_2^- amide
   3. OH^- hydroxide
   4. AsO_3^{3-} arsenite
   5. ClO_2^- chlorite
   6. ClO^- hypochlorite
   7. $S_2O_4^{2-}$ dithionite
   8. SO_3^{2-} trioxidosulfate(2−) **or** sulfite
   9. NO_2^- dioxidonitrate(1−) **or** nitrite
   10. $[Fe(CO)_4]^{2-}$ tetracarbonylferrate(2−)
   11. $[Cr(NCS)_4(NH_3)_2]^-$ diamminetetrathiocyanatochromate(1−)

Examples 8–11 show names developed using additive nomenclature. However, note the method used to express the charge on an ion, either positive or negative. This method is used whichever way the name was derived, by an arabic numeral in parentheses followed by + or – depending on the sign of the charge, and placed as a suffix. Indicating of charges by symbols such as –2 or +3 is always incorrect, since the convention which IUPAC follows is that these represent simple numbers and can never imply numbers of charges. In addition, IUPAC does not approve the use of repeated signs, such as $--$ and $+++$, to represent charges correctly designated 2– and 3+.

As noted for Examples 15–17 in Section 5.3 above, homopolyatomic anions take systematic names of the kind exemplified below.

Examples

12.	O_2^-	dioxide(1–)
13.	I_3^-	triiodide(1–)
14.	Pb_9^{4-}	nonaplumbide(4–)

Where a group of anions is cited in a name, the order of citation is always alphabetical. The use of multiplicative prefixes does not affect this order unless the prefix is part of the anion name. For example, 'tris(iodide)' meaning three anions I^- is ordered under 'i', but '(triiodide)(1–)' is ordered under 't'. Multiplicative prefixes are listed in Table P6. If two successive multiplicative prefixes are required, the alternative prefixes shown in Table P6 are recommended to be employed for the second prefix in the manner shown: $Ca(I_3)_2$, calcium bis(triiodide). This is to avoid any possible confusion.

Oxidation state can be indicated in a name using a Roman numeral in parentheses as a suffix to the name of the moiety being qualified. Unless otherwise indicated by a minus sign, an oxidation state is always taken to be positive. All these devices, charge and oxidation state indication, multiplicative prefixes and also the hierarchy of enclosure marks, are exemplified below. It is evident that there is often more than one method for generating these names to convey stoichiometric information. Employing them all would create redundancies and this is to be avoided. In particular, qualification by both charge and oxidation number is not recommended. The Examples below show a selection of these usages.

Examples

15.	UO_2^{2+}	dioxidouranium(2+)
16.	Na^-	sodide(1–)
17.	PO_4^{3-}	tetraoxidophosphate(V) **or** phosphate
18.	N_2O	nitrogen(I) oxide
19.	Fe_3O_4	iron(II) diiron(III) tetraoxide
20.	SF_6	sulfur(VI) fluoride
21.	UO_2SO_4	dioxidouranium(VI) tetraoxidosulfate(VI) **or** dioxidouranium(VI) sulfate
22.	$(UO_2)_2SO_4$	bis[dioxidouranium(V)] tetraoxidosulfate(VI) **or** bis[dioxidouranium(V)] sulfate

23. Hg_2Cl_2	dimercury(I) chloride **or** dimercury(2+) chloride
24. $Fe_2(SO_4)_3$	iron(3+) sulfate **or** iron(III) sulfate **or** diiron tris(sulfate)

Anions obtained formally by the loss of a hydron or hydrons (see Section 2.2) from a parent hydride (see Table P8 and Chapter 6 for a list of parent hydride names) are conveniently named by the methods of substitutive nomenclature, which are discussed in detail in Chapter 6. Note that hydron is the recommended name for the normal isotopic mixture of protons, deuterons and tritons found in, say, water; the designation proton for this mixture is clearly a misnomer, though it is still widely used (Section 2.2).

Examples

25. CH_3^-	methanide
26. NH_2^-	amide **or** azanide (based on the name azane for NH_3)
27. PH^{2-}	phosphanediide
28. SiH_3^-	silanide

Anions can also be formed by the loss of hydrons from molecules of acids. Where all the available hydrons have been lost the name of the acid is modified as shown below, Examples 29–31. If not all the hydrons are lost, the modification of the name of the parent acid is rather more complex, as shown in Examples 32 and 33 below. Additive names, some of which are shown in Examples above, are often preferred for such species. For more examples of these names, see Table 5.1 and a more complete list in Table IR-8.1 of the *Nomenclature of Inorganic Chemistry, Recommendations 2005*. Note that many trivial names are still allowed, even where they

Table 5.1 Names for some selected oxoacids and derivatives

Formula	Acceptable Common Name	Systematic Additive Name (see Chapter 7)
CO_3^{2-}	carbonate	trioxidocarbonate(2–)
HCO_3^-	hydrogencarbonate	hydroxidodioxidocarbonate(1–)
H_2CO_3	carbonic acid	dihydroxidooxidocarbon
PO_4^{3-}	phosphate	tetraoxidophosphate(3–)
HPO_4^{2-}	hydrogenphosphate	hydroxidotrioxidophosphate(2–)
$H_2PO_4^-$	dihydrogenphosphate	dihydroxidodioxidophosphate(1–)
H_3PO_4	phosphoric acid	trihydroxidooxidophosphorus
PO_3^{3-}	phosphite	trioxidophosphate(3–)
HPO_3^{2-}	hydrogenphosphite	hydroxidodioxidophosphate(2–)
$H_2PO_3^-$	dihydrogenphosphite	dihydroxidooxidophosphate(1–)
H_3PO_3	trihydrogenphosphite	trihydroxidophosphorus
$H_4P_2O_7$	diphosphoric acid	μ-oxido-bis(dihydroxidooxidophosphorus)
H_2SO_4	sulfuric acid	dihydroxidodioxidosulfur
H_2SO_3	sulfurous acid	dihydroxidooxidosulfur
H_4SiO_4	silicic acid	tetrahydroxidosilicon
HOCN	cyanic acid	hydroxidonitridocarbon
HNCO	isocyanic acid	(hydridonitrato)oxidocarbon
$H_2S_2O_6$	dithionic acid	bis(hydroxidodioxidosulfur)(*S—S*)
HClO	hypochlorous acid	chloridohydridooxygen
$HClO_4$	perchloric acid	hydroxidotrioxidochlorine

use the endings 'ous' and 'ic', which are no longer recommended for the names of element cations. The systematic names in Table 5.1 may contain more information, but they are often longer.

Examples

29. H_2SO_4 sulfuric acid $\rightarrow SO_4^{2-}$, sulfate
30. H_3PO_4 phosphoric acid $\rightarrow PO_4^{3-}$, phosphate
31. CH_3COOH ethanoic acid **or** acetic acid (trivial name) $\rightarrow CH_3COO^-$, ethanoate **or** acetate (trivial name)
32. HSO_4^- hydrogensulfate
33. $H_2PO_4^-$ dihydrogenphosphate

The termination 'ate' is also used when hydrons are formally subtracted from an OH group in alcohols, phenols, *etc*. For more details see Chapter 6 on substitutive nomenclature.

Examples

34. CH_3O^- methanolate
35. $C_6H_5O^-$ phenolate
36. $C_6H_5S^-$ benzenethiolate

An alternative way of envisaging anion formation in certain cases is by reaction of a neutral molecule or atom with hydride ions, H^-. The 'ate' termination is also used in such cases. There are two ways to envisage such a process: the addition of $(n + 1)$ hydride ions to a central cation, M^{n+}, to form an anion, MH_{n+1}^-, or the addition of a single hydride ion to a parent hydride, MH_n. Examples 37–39 exemplify the first method, and are examples of coordination nomenclature.

Examples

37. BH_4^- $= B^{3+} + 4H^- \rightarrow$ tetrahydridoborate
38. PH_6^- $= P^{5+} + 6H^- \rightarrow$ hexahydridophosphate
39. BCl_3H^- $= B^{3+} + 3Cl^- + H^- \rightarrow$ trichloridohydridoborate

Note the use of the systematic hydrido for the ligand H^- rather than the name hydro, since this ligand name is now to be treated in a way consistent with those of other monoatomic anionic ligands (as in Example 39).

Substitutive nomenclature rules suggest the names boranuide for BH_4^- and λ^5-phosphanuide for PH_6^-. For further information on these names and on the use of λ to indicate the bonding number in such circumstances, see Chapter 6.

5.5   NAMES OF POLYATOMIC CATIONS

The names of monatomic cations have been mentioned briefly above (Section 5.2). These and the names, including the charges, of polyatomic cations can be designated quite simply.

Examples

1. Na^+ sodium(1+)
2. U^{6+} uranium(6+)

3. I^+ iodine(1+)
4. O_2^+ dioxygen(1+)
5. Bi_5^{4+} pentabismuth(4+)
6. Hg_2^{2+} dimercury(2+)

In running text one might wish to write phrases such as 'dioxygen(1+) ion' or 'sodium(I) cation', but the words 'ion' and 'cation' are not parts of the names of the species discussed.

Names for some cations can also be obtained by formal addition of a hydron to a binary hydride. In such cases, a formalism of substitutive nomenclature (Chapter 6) is used; the final e of the parent name from which the cation may be considered to be derived is replaced by the suffix 'ium'. For the names of appropriate parent hydrides, see Table P8 and Chapter 6.

Examples

7. H_3S^+ sulfane → sulfanium
8. PH_4^+ phosphane → phosphanium
9. SiH_5^+ silane → silanium

Some variants previously allowed for mononuclear cations of elements of Groups 15, 16 and 17 are no longer recommended, for example phosphonium should now be called phosphanium, sulfonium is now called sulfanium, and iodonium is now called iodanium. These original names followed the pattern based upon ammonium, but are no longer recommended. The names ammonium for NH_4^+ and oxonium for H_3O^+ are still acceptable, together with the systematic azanium and oxidanium, because they are hallowed by long usage.

Derivatives of these hydrides, including organic derivatives, are named using the rules of substitutive nomenclature (Chapter 6), or by using additive nomenclature (Chapter 7), whichever seems the more appropriate.

Examples

10. $[PCl_4]^+$ tetrachlorophosphanium (substitutive)
or tetrachloridophosphorus(1+) (additive)
11. $[P(CH_3)_2Cl_2]^+$ dichlorodimethylphosphanium (substitutive)
or dichloridodimethylphosphorus(1+) (additive)

Where a cation can clearly be regarded as a coordination complex, as in Example 12, additive nomenclature should be applied. Such applications are discussed in detail in Chapter 7.

Example

12. $[CoCl(NH_3)_5]^{2+}$ pentaamminechloridocobalt(2+)

There are very few cases where trivial names for important cations are still allowed. The ion NO^+ was formerly called the nitrosyl cation, but the recommended name is now oxidonitrogen(1+). Similarly, NO_2^+ was formerly called the nitryl cation, but it is now named dioxidonitrogen(1+). However, hydroxylium is still allowed for OH^+, though the systematic oxidanylium is recommended.

5.6 APPLICATIONS OF ANION AND CATION NAMES

5.6.1 Simple salts

The names of some salts have naturally already been referred to in places in the foregoing text. Here the naming process is summarised. Salts, whether inorganic or organic, have binary names that are often indistinguishable in kind from binary names for compounds which are not salts, and which are obtained by classifying the constituents as electropositive and electronegative species. A binary name does not necessarily imply a salt-like character.

The names of cations in the name of a salt always precede the names of anions in English and are always separated by spaces, without exception. The order of citation is alphabetical, with the possible exception of hydrogen. If the hydrogen is present as a cation, hydrogen(1+), its name is cited last amongst the cations. If the hydrogen is part of an anion, as in HSO_4^- or HCO_3^-, it is cited as part of the anion name, to which it is joined without a space, as shown in Table 5.1. The anion names are then also ordered alphabetically within their own group in the complete compound name.

Examples

1. $NaHCO_3$	sodium hydrogencarbonate
2. K_2HPO_4	dipotassium hydrogenphosphate
3. $KMgF_3$	**m**agnesium **p**otassium fluoride
4. $Na(UO_2)_3[Zn(H_2O)_6](CH_3CO_2)_9$	tris(**d**ioxidouranium) **h**exaaquazinc sodium nonaethanoate
5. $Ca_5F(PO_4)_3$	pentacalcium **f**luoride tris(**p**hosphate)

In Example 3, trifluoride rather than fluoride would be equally acceptable, but the multiplicative prefix is not absolutely necessary since the conventional but non-explicit cation charges necessarily imply three anions. In Example 4, nonaacetate would be just as acceptable as nonaethanoate.

5.6.2 Double and multiple salts

Double and multiple salts are named simply using the binary system exemplified above. This is logical because two (or more) individual salts are not distinguishable in the real material, neither in the solid state nor in solution.

Example

6. $NaCl \cdot NaF \cdot 2Na_2SO_4$	hexasodium chloride fluoride bis(sulfate)

However, such compounds may sometimes be regarded as (formal) addition compounds and named as such (see Section 5.6.3).

5.6.3 **Addition compounds**

Example 12 of Section 5.3 above is a member of a class of compound termed addition compounds. This covers donor-acceptor complexes as well as a variety of lattice compounds, often of uncertain structure. These can all be named by citing the names of the constituent compounds and then indicating their proportions as discussed in Section 5.3. Hydrates and salt hydrates constitute a large class of compound that can be represented by this means. In Example 7 below an equally satisfactory name is ammonium magnesium phosphate hexahydrate. This method of naming is not often applied to other types of hydrate and is rarely used with other kinds of solvate. The bold initial letters in Example 7 below are to emphasise the alphabetical ordering employed, and are not to be reproduced in normal use.

Examples

7. $MgNH_4PO_4 \cdot 6H_2O$	**a**mmonium **m**agnesium **p**hosphate—water (1/6)
8. $3CdSO_4 \cdot 8H_2O$	cadmium sulfate—water (3/8)
9. $CaCl_2 \cdot 8NH_3$	calcium chloride—ammonia (1/8)
10. $BiCl_3 \cdot 3PCl_5$	bismuth(III) chloride—phosphorus(V) chloride (1/3)
11. $Al_2(SO_4)_3 \cdot K_2SO_4 \cdot 24H_2O$	aluminium sulfate—potassium sulfate—water (1/1/24)
12. $NaCl \cdot NaF \cdot 2Na_2SO_4$	sodium chloride—sodium fluoride—sodium sulfate (1/1/2)

Example 11 could equally be named dialuminium dipotassium tetrakis(sulfate) tetracosahydrate, though specification of the degree of hydration might present problems to those unfamiliar with the higher multiplicative prefixes. The compound cited in Example 12 is the same as that in Example 6, but here it is named as an addition compound rather than as a multiple salt.

5.6.4 **Names of miscellaneous groups and radicals**

The names of groups that can be considered as substituents in organic compounds or sometimes as ligands on metals are often the same as those of the corresponding radicals. According to the rules of substitutive nomenclature, names of radicals and of the corresponding substituent groups are generally derived from the parent hydride names by modifying the hydride names with the suffix 'yl' and the elision of any final 'e'. Note that for the parent hydrides of the Group 14 elements, abbreviated forms of the radical names are acceptable (methyl for methanyl, silyl for silanyl, *etc.*).

Examples

12. $SiH_3{}^{\bullet}$	silyl
13. $SnCl_3{}^{\bullet}$	trichlorostannyl
14. $BH_2{}^{\bullet}$	boranyl
15. $CH_3{}^{\bullet}$	methyl

5.7   A CAVEAT FOR THE UNWARY: DIFFERENT NAMES
        FOR A SINGLE COMPOUND

The discussion above presents several alternative ways of constructing names, in part because organic chemists have traditionally used a different methodology to that used by inorganic chemists. Different chemists may make different choices when naming the same compound, so it is important to be flexible when attempting to interpret a name. Nevertheless, one should always be consistent in the use of nomenclature within work being drafted. However, many writers of articles in the literature have not been consistent, so that different kinds of nomenclature may be present, even in different parts of one name. This is sometimes unavoidable, such as when deriving substitutive names to be used for organic compounds as ligands in additive (coordination) names. Examples 1–3 below indicate different approaches to names of some simple compounds. These are selected examples of such names, not a comprehensive list.

In addition, some usages are no longer approved according to the latest nomenclature decisions, but they may be common in the older literature and are certain to fall out of use only slowly. For example, until recently chloride ligands were named 'chloro' within additive names. The currently recommended 'chlorido' is systematic and formed by the same process used to derive the names of other anionic ligands. This provides for a clearer differentiation between additive and substitutive nomenclatures. In the latter, 'chloro' is still used for the characteristic group Cl–. Related anions (oxide, bromide, *etc.*) are treated similarly. The use in English of a different name stem other than that of the element name to name some simple derivatives of that element has also been changed in certain cases, to produce greater consistency. For example, sodium gives rise to an ion Na⁻, now to be called sodide rather than natride as previously. The names used in other languages are a matter for those who practise them.

When elements carrying no ligands themselves are regarded as simple ligands on another central atom, the element name is modified by adding the suffix 'ido' to the stem of the name, as listed for selected cases in Table P7 which is a modified version of Table IX of the *Nomenclature of Inorganic Chemistry, Recommendations 2005*. However, if the element itself is the central atom of a ligand or substituent group, the name of the central atom must be modified by the appropriate rules for substitutive or additive nomenclature (Chapters 6–8), and then adapted to yield the ligand name.

Examples
1. $COCl_2$ dichloridooxidocarbon (This is an additive name. Carbonyl is still used for the ligand CO. In substitutive nomenclature carbonyl is the name for the group >CO and the word is found in some more complex names. $COCl_2$ may also be called by a functional class name derived from carbonic acid, carbonyl dichloride or carbonyl chloride. In this case, the multiplicative prefix may be omitted, as it is often assumed to be implicit. A trivial name for $COCl_2$ is phosgene.)
2. SO_2Cl_2 dichloridodioxidosulfur (This is also an additive name. A functional class name for SO_2Cl_2 is sulfuryl dichloride, derived from the name

sulfuric acid. As may be the practice with some compositional names, the multiplicative prefix may be omitted if it is assumed to be implicit, *viz.*, sulfuryl chloride, and the name sulfuryl chloride for SO_2Cl_2 is still approved, as is thionyl chloride for $SOCl_2$. The latter may also be named additively, with less reliance on memory, dichloridooxidosulfur. The names sulfinyl and sulfonyl for $>SO$ and $>SO_2$, respectively, apply to substituent groups).

3. NOCl   chloridooxidonitrogen (This is an additive name. The name nitrosyl chloride is not cited in the *Nomenclature of Inorganic Chemistry, Recommendations 2005*, though it was in the 1990 edition, but neither has it been deemed unacceptable. Unbound NO is best named nitrogen(II) oxide or nitrogen monoxide, or some equivalent variant. When NO is a ligand it is called nitrosyl.)

In nomenclature, certain groups, such as –OH (hydroxy, or systematically oxidanyl) have trivial names that are still in general use. Some similar names have been used in the foregoing text and further examples are listed in Table P7. However, note that the use of many traditional names for groups, such as chromyl for CrO_2 and uranyl for UO_2, is no longer recommended.

5.8   SUMMARY

This Chapter has presented examples of both trivial and systematic names, which reflect the current usage and state of development of nomenclature. In the following Chapters the systems recommended for substitutive, additive and polymer nomenclatures are described in more detail. Table P7 lists numerous examples of names for molecules, anions, cations and groups, and this should be consulted when name construction is attempted.

6 Substitutive and Functional Class Nomenclature

6.1 INTRODUCTION

Organic nomenclature is a vast subject, and this text is concerned with basic principles. The nomenclature is complicated and it is also developing continuously. The most recent changes may be viewed at http://www.iupac.org and at http://www.chem.qmul.ac.uk/iupac/. **Substitutive nomenclature** was developed using the concepts that governed the development of organic chemistry. However, in nomenclature the terms **substitution** and **substitutive** are used in a very restricted sense: only hydrogen atoms can be substituted (or exchanged) with other atoms or groups of atoms. Thus a parent hydride must always be the starting point of a substitution operation. For instance, the molecules CH_3–Cl and CH_3–OH are always considered to be derived from the parent hydride CH_3–H, methane. When atoms other than hydrogen are exchanged, the operation is instead called **replacement** and the resultant nomenclature is called **replacement nomenclature** (see Section 6.3.3 for examples of this method).

A **substitutive name** consists of the name of a parent hydride to which prefixes and suffixes are attached as necessary, following the general pattern:

prefixes — name of parent hydride — suffixes

By definition, an organic molecule must contain at least one carbon atom. A given organic molecule is composed of a carbon skeleton and, except for parent hydrides, (characteristic or functional) groups. A name matches a structure if the name of the parent hydride corresponds to the skeleton, while the prefixes and suffixes represent the characteristic groups and other structural characteristics of the molecule, such as geometry. An organometallic molecule is defined as an entity containing a direct bond between a carbon atom and a metal atom. In such molecules, the skeleton may be composed entirely or partly of atoms other than carbon. Such compounds are also discussed in Chapter 8.

The material contained in this Chapter encompasses a considerable amount of organic chemistry. Teachers should be careful not to overwhelm beginning students, but the subject matter of this Chapter will be sufficient for the needs of most, except actual or aspiring organic chemistry specialists.

Principles of Chemical Nomenclature: A Guide to IUPAC Recommendations
Edited by G. J. Leigh
© International Union of Pure and Applied Chemistry 2011
Published by the Royal Society of Chemistry, www.rsc.org

CHAPTER 6

6.2 ALKANES AND THE BASIC APPROACH TO SUBSTITUTIVE NAMES

6.2.1 General

Alkanes are acyclic hydrocarbons of general formula C_nH_{2n+2}. The carbon atoms are arranged in **chains** that may be **branched** or **unbranched**. Chains composed of $-CH_2-$ groups with two terminal $-CH_3$ groups are called continuous or unbranched. If they contain more than two terminal $-CH_3$ groups they are branched. In such cases, at least one carbon atom must be joined by single bonds to at least three other carbon atoms.

Examples

1. $CH_3-CH_2-CH_2-CH_3$

2. $\begin{array}{c} CH_3 \\ \diagdown \\ CH-CH_2-CH_3 \\ \diagup \\ CH_3 \end{array}$

an unbranched alkane   a branched alkane

6.2.2 Unbranched alkanes

Unbranched or straight-chain alkanes were formerly called 'normal'. Although the names of the first four members of the homologous series of unbranched alkanes, general formula C_nH_{2n+2}, methane, ethane, propane, and butane, are trivial and not systematic, they are **retained names**, that is, they are still acceptable, even though they are not systematic. They were coined more than 100 years ago, officially recognised by the Geneva Conference in 1892, and have been used ever since. There are no alternative names for them.

Examples
3. methane CH_4 4. propane $CH_3-CH_2-CH_3$
5. ethane CH_3-CH_3 6. butane $CH_3-CH_2-CH_2-CH_3$

However, higher members of the series are named systematically by combining the ending 'ane', characteristic of the first four members and implying complete saturation with hydrogen atoms of all valences other than those in single bonds between carbon atoms. A multiplicative prefix of the series 'penta', 'hexa', *etc.* of Table P6 indicates the number of carbon atoms constituting the chain. The letter 'a', which ends the multiplicative prefix, is elided.

Example
7. pent(a) + ane = pentane $CH_3-[CH_2]_3-CH_3$

The names of unbranched alkanes are of the utmost importance, because these alkanes are the parent hydrides used to name all **aliphatic molecules**, *i.e.*, molecules having a carbon-chain skeleton.

6.2.3   **Numbering of unbranched alkanes**

An unbranched chain is numbered from one end to the other using arabic numerals. The direction of numbering is chosen as to give the lowest numbers (locants) to carbon atoms carrying any **substituent groups**, which are the entities attached to the chain (see Section 6.2.7).

Example

8. $\underset{1}{CH_3}-\underset{2}{CH_2}-\underset{3}{CH_2}-\underset{4}{CH_2}-\underset{5}{CH_3}$

6.2.4   **Branched alkanes**

Branched-chain alkanes can be considered to be constituted of a principal chain and side-chains. They are named by using the precise set of operations listed below.
1. Selection of the principal chain, which will serve as the parent hydride.
2. Identification and naming of side-chains, which will be cited in the name as prefixes.
3. Determination of the position of side-chains on the principal chain and selection of locants using the rule of lowest locants, that is, the carbon atoms carrying the side-chains receive the lowest possible set of locants (see Section 6.2.7).
4. Selection of the appropriate multiplicative prefixes.
5. Construction of the full name.

The following example illustrates the step-by-step construction of the name of the branched alkane shown.

Example

9.
$$\underset{1}{CH_3}-\underset{2}{CH_2}-\underset{3}{\overset{\overset{\displaystyle CH_3}{|}}{CH}}-\underset{4}{CH_2}-\underset{5}{CH_2}-\underset{6}{CH_2}-\underset{7}{CH_3}$$

Using the rules of substitutive nomenclature and in the absence of a characteristic group which would have to be cited in the name as a suffix, the construction of the name begins by selecting the longest chain, which in this case has seven carbon atoms. The disallowed alternatives have four or six. The parent hydride is therefore heptane and there is one carbon atom in the side-chain. This is a methyl group according to Section 6.2.5, leading to a partial name, 'methylheptane'. Multiplicative prefixes are not needed in this Example, as the prefix 'mono' is never used in a substitutive name. Finally, the locant '3' is placed immediately in front of the 'methyl', the substituent name it qualifies. The alternative '5' is larger, and is therefore rejected. Locants are separated from other parts of the name by hyphens. Thus the full name is 3-methylheptane.

The three trivial names in Examples 10–12 are still retained, but only for the unsubstituted hydrocarbons. Derivatives should always be named using systematic procedures.

Examples

10. $(CH_3)_2CH-CH_3$ isobutane
11. $(CH_3)_2CH-CH_2-CH_3$ isopentane
12. $(CH_3)_4C$ neopentane

6.2.5 **Names of substituent groups**

Removal of a single hydrogen atom from an alkane produces a monovalent group. When the hydrogen atom is removed from the terminal CH_3 group of a parent hydride which is an unbranched alkane, traditionally the product is called an **alkyl group**. These are named by replacing the ending 'ane' in the name of the parent hydride by the suffix 'yl'. By definition, the hydrogen atom is removed from a carbon atom bearing the locant 1, though this locant is never cited. Unbranched groups resulting from the subtraction of one hydrogen atom from a non-terminal position are called **alkanyl groups** (**not** alkyl groups) and are named by adding the suffix 'yl' to the name of the parent hydride with the elision of the final letter 'e' of the alkane name. A locant is always used to denote the position of this free valence. The locant is assigned so as to be as low as possible.

Examples

13. $-CH_3$ methyl
14. $-CH_2-CH_3$ ethyl
15. $-CH_2-CH_2-CH_3$ propyl
16. $-CH_2-CH_2-CH_2-CH_3$ butyl
17. $-CH_2-[CH_2]_8-CH_3$ decyl

18. $\underset{3}{CH_3}-\underset{2}{\overset{|}{CH}}-\underset{1}{CH_3}$ propan-2-yl

19. $\underset{4}{CH_3}-\underset{3}{CH_2}-\underset{2}{\overset{|}{CH}}-\underset{1}{CH_3}$ butan-2-yl
(**not** butan-3-yl)

20. $\underset{10}{CH_3}-[CH_2]_6-\underset{3}{\overset{|}{CH}}-CH_2-\underset{1}{CH_3}$ decan-3-yl
(**not** decan-8-yl)

Branched alkyl or alkanyl groups are named by prefixing the names of the side-chains to that of the longest unbranched alkyl or alkanyl group.

Examples

21. $\underset{5}{CH_3}-\underset{4}{CH_2}-\underset{3}{CH}(CH_3)-\underset{2}{CH_2}-\underset{1}{CH_2}-$ 3-methylpentyl

22. $\underset{5}{CH_3}-\underset{4}{CH_2}-\underset{3}{\overset{\overset{\displaystyle CH_3}{|}}{CH}}-\underset{2}{\overset{|}{CH}}-\underset{1}{CH_3}$ 3-methylpentan-2-yl

23. $\underset{2}{CH_3}-\underset{1}{\overset{|}{CH}}-CH_3$ 1-methylethyl (**or** propan-2-yl)

24. $\underset{3}{CH_3}-\underset{2}{CH_2}-\underset{1}{\overset{|}{CH}}-CH_3$ 1-methylpropyl (**or** butan-2-yl)

For the unsubstituted groups in Examples 25–31, the non-systematic names are still used, and these are also retained names. If there are substituents within these groups, systematic procedures based on the names of the appropriate unbranched chain must be followed to derive the names.

Examples
25. $(CH_3)_2CH-$ isopropyl
26. $(CH_3)_2CH-CH_2-$ isobutyl
27. $CH_3-CH_2-CH(CH_3)-$ *sec*-butyl
28. $(CH_3)_3C-$ *tert*-butyl
29. $(CH_3)_2CH-CH_2-CH_2-$ isopentyl
30. $CH_3-CH_2-C(CH_3)_2-$ *tert*-pentyl
31. $(CH_3)_3C-CH_2-$ neopentyl

The groups attached to a principal chain, the **substituent groups**, may be simple or complex. Simple substituents are formed directly from a parent hydride. When a simple substituent is itself substituted, it is termed a complex substituent. Straight-chain alkyl and alkanyl groups are simple substituents; branched alkyl and alkanyl groups are complex substituents. However, as exceptions, the names isopropyl, isobutyl, *sec*-butyl, *tert*-butyl, isopentyl, *tert*-pentyl, and neopentyl are treated as simple substituents.

6.2.6   Multiplicative prefixes

Multiplicative prefixes (Table P6) are used when more than one substituent of a given kind is present in a compound or group. The name of the substituent is cited as a prefix, and one of the two sets of multiplicative prefixes is used, depending on whether the substituent is simple or complex.

Basic multiplicative prefixes 'di', 'tri', 'tetra', *etc.*, are used with the names of simple substituents and retained names. The alternative modified prefixes in Table P6 are used with complex substituents: 'bis', 'tris', 'tetrakis', *etc.*; from 'tetrakis' onward, the ending 'kis' is attached to the basic multiplicative prefix, giving 'pentakis', 'hexakis', *etc.* (compare the use of these multiplicative prefixes in coordination nomenclature, Chapter 7). In names, parentheses are always placed around the substituent names after the multiplicative prefixes 'bis', 'tris', *etc.* When complex substituents are further substituted, nested enclosing marks are used, parentheses (), square brackets [], and braces {}, in the order {[()]}.

Examples
32. 3,3-dimethylpentane
33. 5,5-bis(1,2-dimethylpropyl)nonane **or**
    5,5-bis(3-methylbutan-2-yl)nonane
34. 4,4-diisopropylheptane **or**
    4,4-bis(1-methylethyl)heptane **or**
    4,4-di(propan-2-yl)heptane

6.2.7   Lowest locants

Locants are used to indicate the position of the substituents in a compound or group. When there is an option in the construction of names in choosing a set of locants, the invariable rule is that the locants are selected so that the set of locants reading in order from left to right has the lowest possible values as determined by comparing alternative possible sets. When compared term-by-term with other locant sets, each in order of increasing magnitude, a set of lowest locants reading from left to right has the lowest term at the first point of difference, for instance, 2,3,6,8 is lower than 3,4,6,8 or 2,4,5,7.

Examples

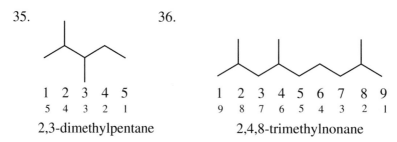

35.

1	2	3	4	5
5	4	3	2	1

2,3-dimethylpentane

36.

1	2	3	4	5	6	7	8	9
9	8	7	6	5	4	3	2	1

2,4,8-trimethylnonane

In Examples 35 and 36, two sets of locants are possible for the principal chains, depending upon the ends at which the numbering of the chains start. In Example 35, the numbering in larger type is selected, because it gives a locant set of 2,3 for the methyl substituents, rather than the alternative 3,4 which is numerically greater in the sense discussed above. Similarly in Example 36, the numbering in larger type is selected for the principal chain because it gives a locant set for the methyl substituents of 2,4,8, smaller than the alternative 2,6,8.

6.2.8   **Alphabetical order for citation of detachable prefixes**

As discussed above, prefixes are used to name substituents. These are called **detachable** or alphabetised prefixes. A further class of prefix is described as **non-detachable**. For example 'cyclo' in cyclohexyl describes a structure which is different in stoichiometry, C_6H_{11}, from the unmodified hexyl, C_6H_{13}, and it must therefore be an integral part of the name. Non-detachable prefixes are used to modify permanently the name of a parent hydride and thus create a new parent hydride (see also Section 6.3.4).

When constructing a name, detachable prefixes are cited in alphabetical order at the front of the name of the parent hydride. The names are alphabetised by considering the first letter in each name: 'm' in methyl, 'b' in butyl, and 'd' in 1,2-dimethylpropyl, *etc*. In retained names, the first non-italicised letter is used for alphabetisation: 'i' in isopropyl, but 'b' in *tert*-butyl.

The construction of a name starts by attaching the names of the detachable prefixes in alphabetical order to the name of the parent hydride. Then, and only then, necessary multiplicative prefixes are introduced without changing the alphabetical order obtained previously. Finally, the locants are inserted. The bold letters in Examples 37–39 are only to indicate how the ordering was arrived at, and should not be emboldened in normal practice.

Examples

37. 4-**e**thyl-2-**m**ethylhexane
38. 4-**e**thyl-2,2-di**m**ethylhexane
39. 6,6-bis(2,3-**d**imethylbutyl)-3,4-di**m**ethylundecane

In some names with more than one detachable prefix, at first sight the locants can be assigned in more than one way, as with the locants 3 and 5 in Example 40. In such cases, the lowest locant is allocated to the substituent cited first.

Example
40. **3-ethyl-5-methylheptane not 5-ethyl-3-methylheptane**

As a consequence, the general pattern of substitutive names becomes:

$$\begin{Bmatrix}\text{Alphabetical citation} \\ \text{of detachable prefixes}\end{Bmatrix} \quad \begin{Bmatrix}\text{non-detachable} \\ \text{prefixes}\end{Bmatrix} \quad \begin{Bmatrix}\text{name of parent} \\ \text{hydride}\end{Bmatrix} \quad \begin{Bmatrix}\text{suffixes}\end{Bmatrix}$$

6.2.9 Criteria for the selection of the principal chain

Where the identity of the parent is not self-evident, it is necessary to lay down rules for its selection. The criteria for the selection of the principal chain are listed in Table 6.1. The criteria described here differ in detail, particularly with respect to unsaturation and chain length, from those published in the *Nomenclature of Organic Chemistry, Recommendations 1979*, and in the *Guide to IUPAC Nomenclature of Organic Compounds, Recommendations 1993*, but are consistent with the usage employed in the *Nomenclature of Inorganic Chemistry, Recommendations 2005* and that adopted in the current revision of the rules of IUPAC organic nomenclature. The criteria are very general as they deal with saturated and unsaturated molecules and include the use of prefixes and suffixes to characterise all kinds of substituent. In the case of alkanes, neither criterion 1 in Table 6.1 nor criterion 2(d) is relevant (there is no principal characteristic group). Criteria 2(b), 2(c), 2(e) and 2(f) are irrelevant (there is no unsaturation). Criterion 2(a) applies, since the chain should be the longest. The next relevant criterion 2(g) is applied when criterion 2(a) does not permit a definitive choice to be made. Accordingly, the principal chain will be that most substituted among all those of equal length under consideration.

Examples

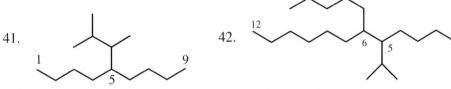

41.

5-(1,2-dimethylpropyl)nonane **or**
5-(3-methylbutan-2-yl)nonane
Criterion 2(a) applies.

42.

5-isopropyl-6-pentyldodecane **or**
5-(1-methylethyl)-6-pentyldodecane **or**
6-pentyl-5-(propan-2-yl)dodecane
Criteria 2(a) and then 2(g) apply.

6.2.10 Generalised 'ane' substitutive nomenclature

Carbon is unique amongst the atoms in the number of hydrides it forms, but the elements near carbon in the Periodic Table have a similar, if more restricted,

Table 6.1 Rules for the selection of a parent structure

When a choice has to be made of a principal parent structure on which a name has to be based, sets of criteria are applied successively, in the given specific order, until all possibilities but one are excluded.

1) Where a choice is necessary, the **senior principal parent structure** has the maximum number of principal characteristic groups cited as a suffix, consistent with the seniority of classes shown in Table 6.5. Thus a dicarboxylic acid is senior to a carboxylic acid. If this is not definitive, then further criteria are applied, as shown below.

2) When a choice of the **principal chain in an acyclic compound** has to be made, the following criteria are applied successively, in the order listed, until all possibilities but one have been excluded. This is then the principal chain.

(a) Chain with maximum length.

(b) Chain with greatest total number of double and triple bonds.

(c) Chain with the greatest number of double bonds.

(d) Chain with lowest locants of principal groups, which are those cited as suffixes.

(e) Chain with the lowest locants for multiple bonds.

(f) Chain with the lowest locants for double bonds.

(g) Chain with the greatest number of substituents cited as prefixes.

(h) Chain with lowest locants for all substituents in the principal chain which are cited as prefixes.

(i) Chain with the substituent that comes earliest in alphabetical order.

3) If the compound contains rings, then any chains are treated as substituents upon the rings. The first five criteria (selected from the long list of criteria for **seniority of rings or ring systems**) are given below.

(a) The most senior ring system is heterocyclic.

(b) The most senior heterocycle contains at least one nitrogen atom.

(c) In the absence of nitrogen, the heterocycle which contains at least one heteroatom of the kind that occurs earlier in the sequence:

$$F > Cl > Br > I > O > S > Se > Te > P > As > Sb > Bi > Si > Ge > Sn > Pb > B > Al > Ga > In > Tl^*$$

(d) The ring system with the greatest number of rings is senior.

(e) If this is not definitive, then the ring or ring system with most ring atoms is selected.

*See also Table 6.3.

propensity to form hydrides, and these can be treated as are the alkanes in substitutive nomenclature. The system of parent hydrides has been extended to all elements of Groups 13, 14, 15, 16, and 17 as indicated in Table P8 for the mononuclear parent hydrides.

Most names listed in Table P8 are formed systematically by joining the ending 'ane' to a stem of an element name, for example: germanium → 'germ' + 'ane' = germane (patterned after methane). The ending 'ane' signifies that the element exhibits its standard bonding number (*i.e.*, the conventional number of electron-pair bonds), 3 for elements of Groups 13 and 15, 4 for elements of Group 14, 2 for elements of Group 16 and 1 for elements of Group 17. Nevertheless, the recommended names in Table P8 are not all formed systematically. In particular, the names oxidane, sulfane, selane, tellane, polane, bismuthane, and alumane are recommended whereas the systematically formed names might be oxane, thiane, selenane, tellurane, polonane, bismane, and aluminane. These, however, are Hantzsch-Widman names (see Tables 6.2 and 6.3 below), used to name ring systems. Indigane is used for InH_3 because the name indane is reserved to designate a fused-ring compound. The names oxidane for water and azane for ammonia are names for use in substitutive nomenclature only. The name phosphine for PH_3 is no longer recommended, phosphane being preferred.

To indicate bonding numbers different from the standard numbers listed above, the **Lambda Convention** is recommended (reference 1). The symbol λ with a superscript indicating the bonding number is added to the name in Table P8, but preceded by a hyphen (see Examples 43–46).

Examples

43. H_2S sulfane
44. H_4S λ^4-sulfane
45. PH_3 phosphane
46. PH_5 λ^5-phosphane

The names of polynuclear hydrides (*i.e.*, compounds consisting of chains) are obtained by prefixing the 'ane' names listed in Table P8 with the appropriate multiplicative prefixes 'di', 'tri', 'tetra', *etc.*

Examples

47. $H_2P–PH_2$ diphosphane
48. $H_3Sn–SnH_2–SnH_3$ tristannane
49. $SiH_3–SiH_2–SiH_2–SiH_3$ tetrasilane

The names of unsaturated compounds are derived by using methods described in Section 6.4). The names dinitrogen and diazyne for N_2 cannot be used in substitutive nomenclature, though both are systematic, since there are no hydrogen atoms to substitute.

Example

50. $H_2N–NH_2$ diazane → HN=NH diazene → N≡N diazyne

The names of substituent groups derived from these hydrides are formed in the manner indicated for alkyl and alkanyl groups in Sections 6.2.5 and 6.4. Silyl, the prefix for $–SiH_3$, germyl, that for $–GeH_3$, stannyl that for $–SnH_3$, and plumbyl that for $-PbH_3$ are similar to alkyl prefixes. All other prefixes take forms similar to those of the alkanyl prefixes.

Examples

51. $-SiH_3$	silyl	52. $-SiH_2-SiH_3$	disilanyl
53. $=GeH_2$	germylidene	54. $-SnH_3$	stannyl
55. $\equiv PbH$	plumbylidyne	56. $-NH-NH_2$	diazanyl

6.3   CYCLIC PARENT HYDRIDES

6.3.1   General

Continuous-chain alkanes are the sole parent hydrides for all compounds that have skeletons composed of chains. Cyclic parent hydrides are more diverse. In nomenclature they are classified according to their structure as **carbocycles** (composed of carbon atoms only) and **heterocycles** (composed of carbon atoms and other atoms such as N, O, and Si). They are also classified as **saturated** and **unsaturated**. Saturated cycles (rings) have the maximum possible number of hydrogen atoms attached to every skeletal atom, as judged by a prespecified valence, whereas unsaturated cycles have fewer hydrogen atoms and consequently multiple bonds between pairs of atoms.

Various degrees of unsaturation are possible, that is, a compound may contain fewer or more double and/or triple bonds. Unsaturation may be **cumulative** (which means there are at least three contiguous carbon atoms joined by double bonds, C=C=C) or **non-cumulative** (which is any other arrangement of double bonds, as in C=C–C=C). In nomenclature, unsaturated cyclic parent hydrides have, by convention, the maximum number of non-cumulative double bonds. They are generically referred to as mancudes, a term derived from the acronym MANCUD, MAximum Number of non-CUmulative Double bonds. The word 'mancude' may be used either as an adjective or as a noun. Four classes of cyclic parent hydride are therefore recognised.

Examples

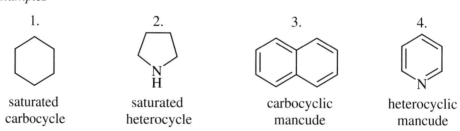

1.	2.	3.	4.
saturated carbocycle	saturated heterocycle	carbocyclic mancude	heterocyclic mancude

A second general feature of the nomenclature of these cyclic compounds is the use of prefixes to modify the names of primary parent hydrides to create new parent hydrides. These prefixes are non-detachable and attached permanently to the name of the primary parent hydride, and they are treated in alphabetical procedures like the non-detachable prefixes in alkane nomenclature (see Section 6.2.8). Among other things, the non-detachable prefixes are used to indicate the conversion of a chain to a cycle (*e.g.*, 'cyclo' as in cyclohexane; see Section 6.2.8), the opening of cycles (*e.g.*, 'seco', as in some natural products; see Chapter 11), the fusion of cycles (*e.g.*, 'benzo', as in 1-benzoxepine; see Section 6.3.4.B) and the replacement of carbon atoms in cycles by heteroatoms, thus transforming carbocycles into heterocycles (*e.g.*, 'phospha', as in phosphacycloundecane; see Section 6.3.3 and Table 6.2).

6.3.2   **Carbocyclic parent hydrides**

Saturated monocyclic carbocycles are generically called cycloalkanes. Individual cycles are named by adding the non-detachable prefix 'cyclo' to the name of the straight-chain alkane having the same number of carbon atoms, for example cyclohexane.

Examples

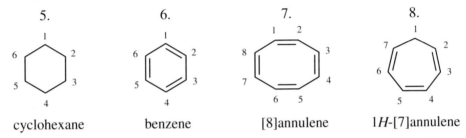

5.	6.	7.	8.
cyclohexane	benzene	[8]annulene	1*H*-[7]annulene

The mancude carbocycle having six carbon atoms is named benzene. Higher mancude homologues having the general formula C_nH_n or C_nH_{n+1} are named [*x*]annulenes, *x* representing the number of carbon atoms in the ring (Example 7). Annulenes with an odd number of carbon atoms are further characterised by the symbol *H* to signal the presence of a unique hydrogen atom called an '**indicated hydrogen**' (Example 8). This symbol is a non-detachable prefix.

The numbering of monocyclic hydrocarbons having no indicated hydrogen depends upon which carbon atom receives the locant '1'. Indicated hydrogen must receive the locant '1', and in such cases the numbering is thus fixed. If other features requiring a prefix to the name are also present, then a choice as to relative seniority has to be made.

6.3.3   **Heteromonocyclic parent hydrides**

These compounds form a large and diverse group. The names of the parent hydrides are usually formed systematically. However, some 50 trivial names are retained and used in preference to their systematic counterparts.

Examples

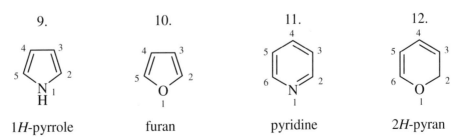

9.	10.	11.	12.
1*H*-pyrrole	furan	pyridine	2*H*-pyran

The two most important methods for systematically naming heteromonocyclic parent hydrides are the extended **Hantzsch-Widman system** and **replacement nomenclature**.

The extended Hantzsch-Widman system is used to name monocycles with saturated and mancude rings containing between three and ten members. The names are composed of two parts: non-detachable prefixes, indicating the heteroatoms, and a stem indicating the size of the ring. Such prefixes (called 'a' prefixes) are listed in Table 6.2 and the stems are shown in Table 6.3.

For the groups of compounds with six ring atoms and designated 6A, 6B and 6C in Table 6.3, the choice of stem is determined by the replacing atom, the name of which immediately precedes the name of the stem. These names are then formed by eliding the final letter 'a' of the 'a' prefix before it is attached to the stem.

When only one heteroatom is present this receives the locant '1'. In many cases an indicated hydrogen is necessary to describe the structure completely. The presence of more than one heteroatom of any type is indicated by a multiplicative prefix ('di', 'tri', *etc.*). If two or more kinds of heteroatom occur in the same name, the order of citation is the descending order of their appearance in Table 6.2. The numbering starts with the heteroatom cited first in Table 6.2 and the lowest possible locants are given to other heteroatoms.

Examples

13.   14.   15.   16.

oxirane   2*H*-azepine   1,2,4-triazine   1,2-oxathiolane

Replacement nomenclature is used to name heteromonocycles that contain more than ten ring atoms. In developing a replacement name, carbon atoms are regarded as replaced by heteroatoms. The non-detachable 'a' prefixes are used to indicate the replacing element. These are similar to most, but not all, of the 'a' prefixes used in the Hantzsch-Widman system (Table 6.2). Cycloalkane names are the bases for transformation into the name of a heterocycle.

Examples

17.   18.

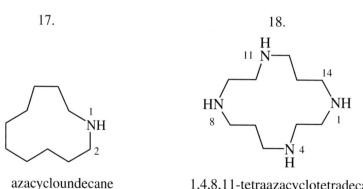

azacycloundecane   1,4,8,11-tetraazacyclotetradecane

Table 6.2 Hantzsch-Widman system prefixes (in decreasing order of seniority) and corresponding replacement nomenclature terms

Element	Bonding number	Prefix	Element	Bonding number	Prefix
fluorine	1	fluora	antimony	3	stiba
chlorine	1	chlora	bismuth	3	bisma
bromine	1	broma	silicon	4	sila
iodine	1	ioda	germanium	4	germa
oxygen	2	oxa	tin	4	stanna
sulfur	2	thia	lead	4	plumba
selenium	2	selena	boron	3	bora
tellurium	2	tellura	aluminium	3	aluma (alumina in replacement nomenclature)
nitrogen	3	aza	gallium	3	galla
phosphorus	3	phospha	indium	3	indiga (inda in replacement nomenclature)
arsenic	3	arsa	thallium	3	thalla

Note that the Hantzsch-Widman prefixes and the 'a' terms used in replacement nomenclature are identical for all the elements listed in Table 6.2 except for those denoting aluminium and indium, which are 'aluma' and 'indiga' in the Hantzsch-Widman system and 'alumina' and 'inda' in replacement nomenclature.

Table 6.3 Stems used in the Hantzsch-Widman system to indicate ring size

Ring Size	Unsaturated	Saturated
3	irene/irine[*]	irane/iridine[*]
4	ete	etane/etidine[*]
5	ole	olane/olidine[*]
6A (O, S, Se, Te, Bi)	ine	ane
6B (N, Si, Ge, Sn, Pb)	ine	inane
6C (F, Cl, Br, I, P, As, Sb, B, Al, Ga, In, Tl)	inine	inane
7	epine	epane
8	ocine	ocane
9	onine	onane
10	ocine	ecane

[*]The stem 'irine' is used in place of 'irene' for rings containing only nitrogen heteroatoms. The stems 'iridine', 'etidine', and 'olidine' are used when nitrogen atoms are present in the ring.

6.3.4   **Polycyclic parent hydrides**

Elementary students may not need the material presented in this Section, but more advanced students may find it useful. Polycyclic parent hydrides are classified as bridged polycycloalkanes (also known as **von Baeyer bridged systems**, from the nomenclature system developed to name them), spiro compounds, fused ring systems, and assemblies of identical cycles. The four classes may be either carbocyclic or heterocyclic. In developing their names, the following rules are used.

A. The non-detachable prefixes 'bicyclo', 'tricyclo', *etc.* and 'spiro' characterise the bridged and spiro systems. No hyphens separate the prefix from the parent name. Numbers in square brackets give necessary information about the length and positions of the branches in these polycyclic systems. The numbering is fixed and obeys precise rules.

Examples

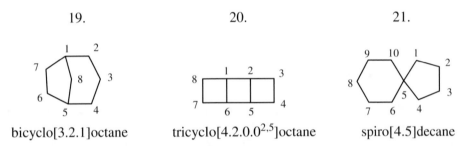

19.	20.	21.
bicyclo[3.2.1]octane	tricyclo[4.2.0.0^{2,5}]octane	spiro[4.5]decane

Heterocyclic systems, which can be regarded as formed by replacement of carbon atoms by heteroatoms in the parent hydrides described above, are named by replacement nomenclature.

Examples

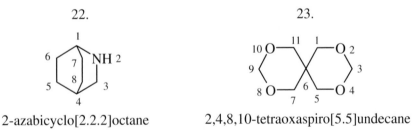

22.	23.
2-azabicyclo[2.2.2]octane	2,4,8,10-tetraoxaspiro[5.5]undecane

B. Fused ring systems are very numerous and diverse and they have both retained names and systematic names. They are systematically named using the structures and names of their smaller components and the concept of *ortho-fusion*, which is a formal operation encountered only in nomenclature. This concept is essential to the naming of larger systems, and is the formation of one bond by the formal fusion (condensation) of two bonds belonging to two different cyclic systems, at least one of them being a mancude cycle or ring system. Many fused polycyclic systems have trivial names that are retained for present use.

Examples

24.

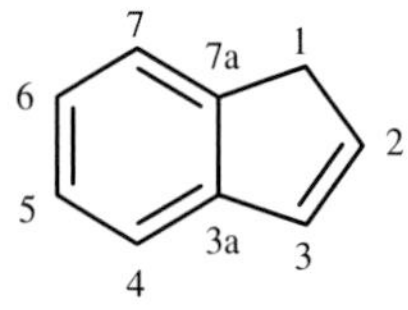

1*H*-indene

25.

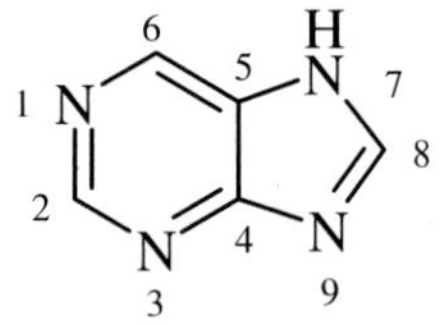

7*H*-purine
(non-systematic numbering)

26.

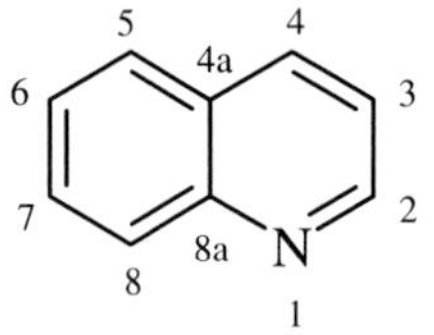

quinoline

27.

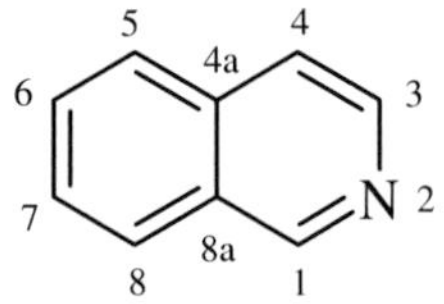

isoquinoline

The numbering system for purines is unique to that class of compound and is not used elsewhere.

Some fused ring systems are named systematically by using a multiplicative prefix before a termination representing a well defined arrangement of rings. For example, the termination 'acene', generalised from the name anthracene, indicates a linear arrangement of fused benzene rings, as in tetracene, Example 29, and pentacene, Example 30.

Examples

28.

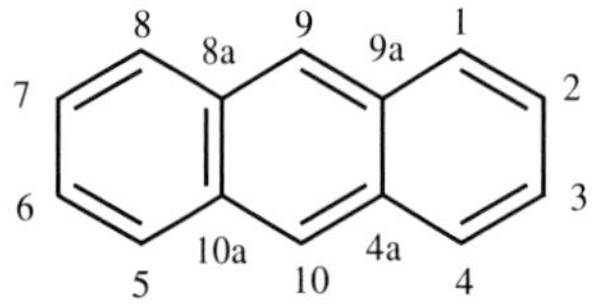

anthracene

29.

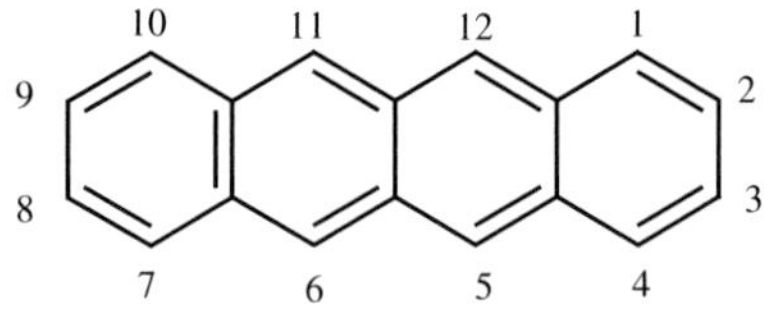

tetracene

30.

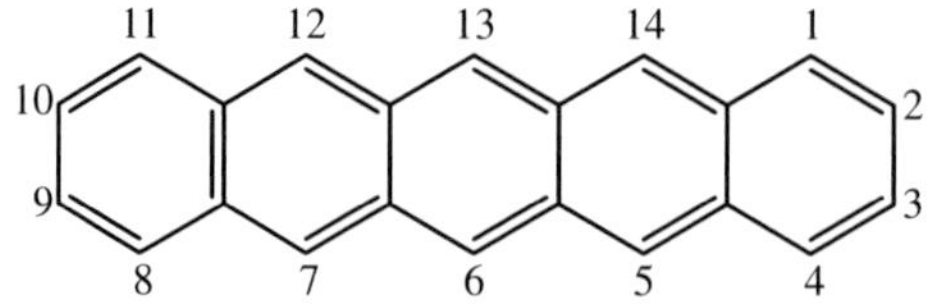

pentacene

If a system does not have a retained name or a name that can be composed systematically as above, and when *ortho*-fusion is possible, the system is named using fusion nomenclature, *i.e.*, by combining the names of the two or more systems that are fused. One system is adjudged to be the senior according to criteria described elsewhere (reference 2) and is taken as a parent hydride, and the other is denoted as a non-detachable prefix. The junction of the two systems is indicated in a specific manner. Instead of numerical locants, italic letters '*a*', '*b*', '*c*', *etc.*, are used to identify bonds in the parent hydride. The final letter 'o' and numerical locants are characteristic of the prefix. Example 31 illustrates the fusion operation and the resulting fusion name.

Examples

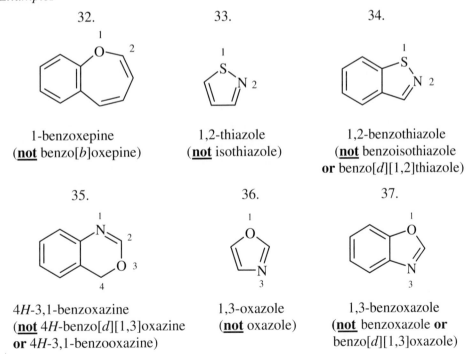

31.

pentalene   +   naphthalene   →   pentaleno[2,1-*a*]naphthalene

***Peri*-fusion** represents another way of formally combining two components, but it is outside the scope of this book (reference 2).

An exception to the general rule of formation of fusion names is used to name benzoheterocycles. A benzene ring fused to a heteromonocycle of five or more members is named by placing the locants indicating the position of heteroatoms at the front of the name, which consists of the prefix 'benzo' followed by a retained name or a Hantzsch-Widman name. Names formed by the general fusion method described above are accepted but are not recommended for these particular compounds.

Examples

32.

1-benzoxepine
(**not** benzo[*b*]oxepine)

33.

1,2-thiazole
(**not** isothiazole)

34.

1,2-benzothiazole
(**not** benzoisothiazole
or benzo[*d*][1,2]thiazole)

35.

4*H*-3,1-benzoxazine
(**not** 4*H*-benzo[*d*][1,3]oxazine
or 4*H*-3,1-benzooxazine)

36.

1,3-oxazole
(**not** oxazole)

37.

1,3-benzoxazole
(**not** benzoxazole **or**
benzo[*d*][1,3]oxazole)

C. Assemblies of identical rings are named by using a unique set of multiplicative prefixes (often termed Latin prefixes) 'bi', 'ter', 'quater', *etc.*, to indicate the number of cycles. Names are formed by using the names of the parent hydrides, except that the name phenyl is used instead of benzene for the mancude six-membered ring. Alternative names can be formed by using the names of the substituent groups to name assemblies of two cycles or ring systems. Locants are primed, double primed, *etc.*, to identify each cycle.

Examples

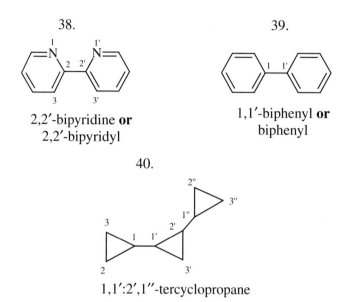

38.

2,2′-bipyridine **or** 2,2′-bipyridyl

39.

1,1′-biphenyl **or** biphenyl

40.

1,1′:2′,1″-tercyclopropane

6.4 NAMES FOR SUBSTANCES WITH VARIOUS DEGREES OF UNSATURATION

6.4.1 General

To denote unsaturation, the ending which indicates saturation in alkanes and mono- and poly-cyclic alkane parent hydrides, 'ane', is changed to 'ene' or 'yne'. These indicate the presence of either a double bond or a triple bond, respectively. If more than one multiple bond of either kind is present, the multiplicative prefixes 'di', 'tri', 'tetra', *etc.*, are used. If locants are needed, they are selected to be the lowest set, and are placed in the name immediately before the endings 'ene' and 'yne'. Note that the name ethylene is no longer approved as a synonym for ethene (Example 1), but it is still allowed for the group –CH_2CH_2– (see Section 6.4.2 and Chapter 9).

Examples

1. $CH_2{=}CH_2$ ethene
2. $CH_2{=}CH{-}CH_2{-}CH_3$ but-1-ene
3. $CH_2{=}CH{-}CH{=}CH_2$ buta-1,3-diene
4. $HC{\equiv}C{-}CH_3$ prop-1-yne
5. $HC{\equiv}CH$ ethyne

6. cyclopentene

7. bicyclo[2.2.2]oct-2-ene

8. cycloocta-1,3,5,7-tetraene

If both double and triple bonds are present together in a structure, locants are still assigned to give the lowest set, considering double and triple bonds together as a group. If this does not allow a resolution, double bonds then receive the lowest locants. The ending 'ene' is cited before the ending 'yne' in a name, but with elision of the final letter 'e'.

Examples

    9. $HC{\equiv}C{-}CH{=}CH{-}CH_3$ pent-3-en-1-yne

   10. $HC{\equiv}C{-}CH{=}CH_2$ but-1-en-3-yne

To indicate the degree of saturation or unsaturation in mancude systems, the non-detachable (but non-alphabetisable) prefixes 'hydro' and 'dehydro' are used to indicate the addition or subtraction of one hydrogen atom. As even numbers of hydrogen atoms are involved when a carbon-carbon single bond becomes a double or triple bond, the multiplicative prefixes 'di', 'tetra', *etc.*, as well as the appropriate locants, are used. The prefix 'dehydro' is never used to indicate the subtraction of hydrogen atoms from saturated heterocycles having Hantzsch-Widman names. These prefixes are placed immediately before the name of the parent hydride.

Examples

11.   12.   13.

1,2-dihydroazulene   1,4-dihydrophosphinine   1,2-didehydrobenzene
                                                    (**not** benzyne)

The name benzyne, often used in the past for didehydrobenzene, is no longer recommended.

The following general pattern for substitutive names follows from applying the rules cited above.

First { alphabetised detachable prefixes } *then* { the **non-detachable prefixes** hydro/dehydro } *then* { name of parent hydride with any structural **non-detachable prefixes,** such as 'cyclo' }

then { **endings** 'ane' or 'ene'/'yne' } *and, finally* { all necessary characteristic **suffixes** }

6.4.2   **Unsaturated and divalent substituents**

Unsaturated monovalent substituents are named systematically by attaching the suffix 'yl' to the name that carries the ending 'ene' or 'yne', with the elision of the final letter 'e' of the name of the alkene or alkyne. Vinyl for $-CH{=}CH_2$, allyl for $-CH_2{-}CH{=}CH_2$, and isopropenyl for $-C(CH_3){=}CH_2$ are amongst those still retained, but isopropenyl may be used only when it is unsubstituted. Otherwise a systematic name must be used.

Examples

14. $-CH_2-CH_2-CH=CH-CH_3$ pent-3-en-1-yl
     1   2   3   4   5

15. $CH_3-CH_2-\overset{|}{C}=CH-CH_3$ pent-3-en-3-yl
     1   2   3   4   5

16. $-CH_2-C{\equiv}CH$ prop-2-yn-1-yl

Names for divalent substituents of the type R–CH= or $R_2C=$ are obtained by appending the suffix 'ylidene' to an appropriate stem. When the two free valences are on different atoms or are of the type $R_2C<$ and not involved in the same double bond, the suffix 'diyl' (di + yl) is used. Retained names include methylene for $-CH_2-$, ethylene for $-CH_2-CH_2-$ and isopropylidene for $=C(CH_3)_2$, but, as usual, only when isopropylidene is unsubstituted.

Examples

17. $=CH_2$ methylidene
18. $=CH-CH_3$ ethylidene
19. $>C=CH_2$ ethene-1,1-diyl
20. $-CH_2-CH_2-CH_2-$ propane-1,3-diyl

A list of divalent groups with two free valences which are not involved in double bonds or are on different atoms, as in Examples 19 and 20, which are often found as constituents of macromolecules (polymers), is to be found in Table P9.

6.4.3   Selection of the principal chain in unsaturated branched acyclic hydrocarbons

To derive the name of a branched unsaturated acyclic hydrocarbon a principal chain must be selected. The general criteria listed in Table 6.1 are applied. Those that are specifically relevant to polyenes and polyynes are criterion 2(a) which is the maximum length, criterion 2(b) which is the maximum number of double and triple bonds considered together, criteria 2(c), 2(e), 2(f) and 2(g), which is the greatest number of substituents cited as prefixes, and criteria 2(h) and 2(i).

Examples

1.

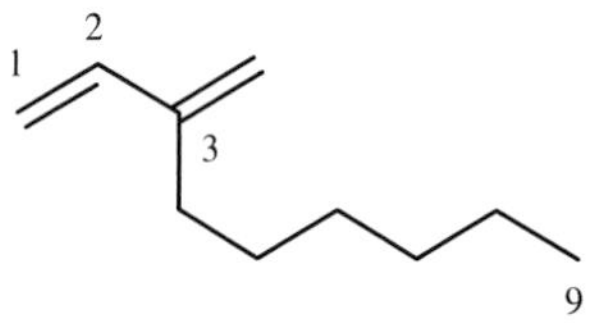

3-methylidenenon-1-ene
Criterion 2(a)

2.

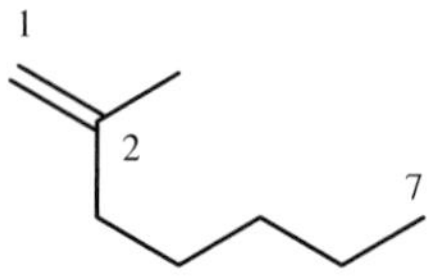

2-methylehept-1-ene
Criteria 2(a) and 2(b)

3.

4-allyl-2-methyloct-1-ene
Criteria 2(a), 2(b) and 2(g)

6.5   ASSEMBLIES OF RINGS, AND OF RINGS WITH CHAINS

Assemblies of different rings are given substitutive names in which one ring is chosen as the parent hydride, and the other is denoted by a prefix. The names of mono- and di-valent cyclic substituents are formed by adding the suffixes 'yl', 'ylidene', 'diyl', *etc.*, as appropriate, to the name of the cycle, with elision of the final letter 'e' from the name of the parent hydride, as discussed in Section 6.2.5 for substituent groups of the alkanyl type. As exceptions, the names phenyl and phenylene are used to describe the mono- and di-valent substituents derived from benzene, and cycloalkanes engender cycloalkyl and cycloalkylidene groups by replacing the ending 'ane' by the suffixes 'yl' and 'ylidene' respectively. Note that wavy lines as shown below

are often used to indicate free valences since formulations

might otherwise be taken to indicate an appended methyl group.

Examples

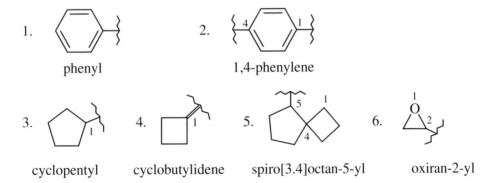

1. phenyl

2. 1,4-phenylene

3. cyclopentyl

4. cyclobutylidene

5. spiro[3.4]octan-5-yl

6. oxiran-2-yl

Some traditional contracted names, such as phenanthryl and thienyl, are retained.

Examples

7.

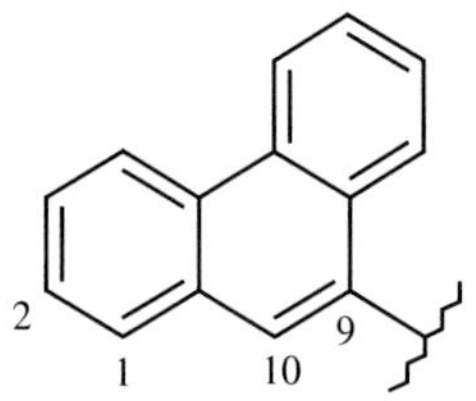

phenanthren-9-yl (similarly the 1-, 2-, 3-, and 4-isomers) **or** 9-phenanthryl
(similarly the 1-, 2-, 3-, and 4-isomers)

8.

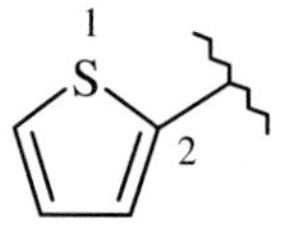

thiophen-2-yl (also the 3-isomer) **or** 2-thienyl (also the 3-isomer)

The principal ring is selected using the criteria listed and summarized in (3) of Table 6.1. For example, heterocycles have seniority over carbocycles, and unsaturated systems have seniority over saturated systems.

Examples

9.

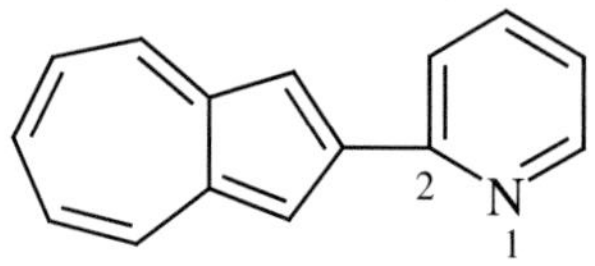

2-(azulen-2-yl)pyridine

10.

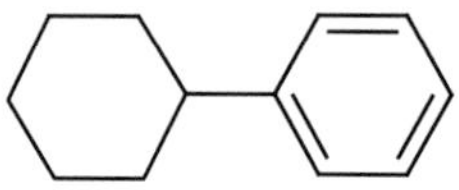

cyclohexylbenzene

Compounds composed of rings and chains are also named using substitutive nomenclature. The ring, whether carbocyclic or heterocyclic, is now always selected as the parent hydride.

Examples

11.

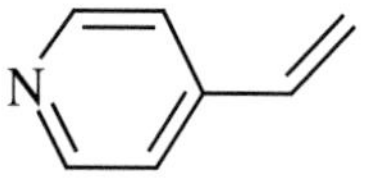

4-vinylpyridine

12.

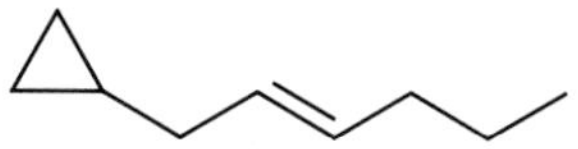

(hex-2-en-1-yl)cyclopropane

Trivial names of substituted benzenes retained for present use include toluene and styrene, but only for derivatives which are substituted in the ring (see Section 6.7).

6.6   NOMENCLATURE OF FUNCTIONALISED PARENT HYDRIDES

6.6.1   The use of suffixes and prefixes

The groups that are typical of the various classes of organic compound, such as –OH in alcohols, >C=O in ketones and –COOH in carboxylic acids are called characteristic groups or functional groups.

Table 6.4 Some trivial names still retained for naming organic compounds

A. Names of functional parent structures, usable with unlimited or limited substitution.

Retained names	Systematic names
Ethers	
anisole (substitution on ring only)	methoxybenzene
Carbonyl compounds	
9,10-anthraquinone and isomers	anthracene-9,10-dione and isomers
1,4-benzoquinone and isomer	cyclohexa-2,5-diene-1,4-dione and isomer
1,2-naphthoquinone and isomers	naphthalene-1,2-dione and isomers
1-isoquinolone and isomers	isoquinolin-1(2H)-one and isomers
2-pyrrolidone and isomer	pyrrolidin-2-one and isomer
2-quinolone and isomers	quinolin-2(1H)-one and isomers
Carboxylic acids	
acrylic acid	prop-2-enoic acid
malonic acid	propanedioic acid
succinic acid	butanedioic acid

B. Names retained only for designating unsubstituted compounds.

Hydroxy compounds	
glycerol	propane-1,2,3-triol
pentaerythritol	2,2-bis(hydroxymethyl)propane-1,3-diol
pinacol	2,3-dimethylbutane-2,3-diol
p-cresol and isomers	4-methylphenol and isomers
thymol	2-isopropyl-5-methylphenol
carvacrol	5-isopropyl-2-methylphenol
pyrocatechol	benzene-1,2-diol
resorcinol	benzene-1,3-diol
hydroquinone	benzene-1,4-diol
picric acid	2,4,6-trinitrophenol
Carboxylic acids	
propionic acid	propanoic acid
butyric acid	butanoic acid
glutaric acid	pentanedioic acid
adipic acid	hexanedioic acid
methacrylic acid	2-methylprop-2-enoic acid
tartaric acid	2,3-dihydroxybutanedioic acid
glyceric acid	2,3-dihydroxypropanoic acid
lactic acid	2-hydroxypropanoic acid
glycolic acid	hydroxyacetic acid
glyoxylic acid	oxoacetic acid
acetoacetic acid	3-oxobutanoic acid
pyruvic acid	2-oxopropanoic acid
citric acid	2-hydroxypropane-1,2,3-tricarboxylic acid
ethylenediaminetetraacetic acid	ethylenedinitrilotetraacetic acid
Amines	
1,2-toluidine and 1,3- and 1,4-isomers	2-methylaniline and 3- and 4-isomers
Hydrocarbons	
isoprene	2-methylbuta-1,3-diene
o-xylene and isomers	1,2-dimethylbenzene and isomers
cumene	isopropylbenzene
o-cymene and isomers	1-isopropyl-2-methylbenzene and isomers
mesitylene	1,3,5-trimethylbenzene
fulvene	5-methylidenecyclopenta-1,3-diene

The names of characteristic groups are cited as suffixes or prefixes when these classes of organic compound are named using substitutive nomenclature (see Table 6.5). If only one characteristic group is present, then its name is cited as a suffix. If more than one type of characteristic group is present, one must be chosen as the principal (senior) characteristic group and its name is then cited as a suffix. The names of all other characteristic groups are cited as detachable (alphabetised) prefixes. However, some characteristic groups are expressed in a name as prefixes only and are never cited as suffixes, *e.g.*, F, 'fluoro' (see Table 6.6). Suffixes and prefixes are listed in Table 6.5. The detailed usage is exemplified in the following discussion.

6.6.2 **Names of characteristic groups always cited as prefixes**

A list of the names of characteristic groups always cited as prefixes is given in Table 6.6. Examples 1 and 2 are a good illustration of the application of the rule of lowest locants and of the use of alphabetical order to assign them.

Examples
1. $ClF_2C–CHBrI$ 2-bromo-1-chloro-1,1-difluoro-2-iodoethane
2. $BrF_2C–CClFI$ 1-bromo-2-chloro-1,1,2-trifluoro-2-iodoethane

In monocyclic hydrocarbons, the locant '1' is omitted if there is only one substituent, but it must be cited for polysubstituted compounds.

Examples

3.

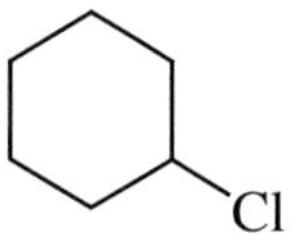

chlorocyclohexane

4.

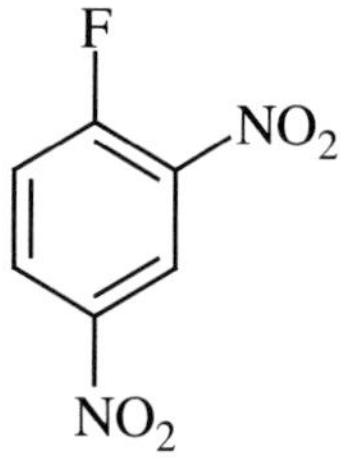

1-fluoro-2,4-dinitrobenzene

5.

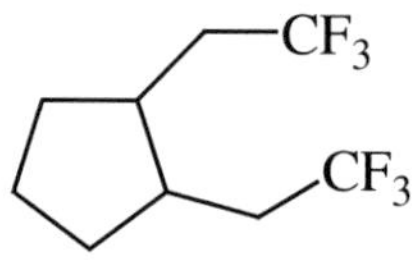

1,2-bis(2,2,2-trifluoroethyl)cyclopentane

When the parent hydride has a fixed numbering, as do polycyclic hydrocarbons and in heterocyclic compounds, selection of lowest possible locants is still the rule.

Table 6.5 Some characteristic groups[*] with names cited as suffixes or prefixes in substitutive nomenclature, presented in decreasing order of seniority of classes

Classes	Suffix	Prefix	Group
Radicals			
Anions	see Table 6.7		
Cations			
Carboxylic acids	oic acid		$-(C)OOH$
	carboxylic acid	carboxy	$-COOH$
Sulfonic acids	sulfonic acid	sulfo	$-SO_2OH$
Esters	see Section 6.11	(R-oxy)/oxo	$-(C)OOR$
		R-oxycarbonyl	$-COOR$
Acyl halides	see Section 6.11	halo/oxo	$-(C)OHal$
		halocarbonyl	$-COHal$
Amides	amide	amino/oxo	$-(C)ONH_2$
	carboxamide	carbamoyl	$-CONH_2$
Nitriles	nitrile		$-(C)N$
	carbonitrile	cyano	$-CN$
Aldehydes	al	oxo	$-(C)HO$
	carbaldehyde	formyl	$-CHO$
Ketones	one	oxo	$=O$
Alcohols, phenols	ol	hydroxy	$-OH$
Amines	amine	amino	$-NH_2$
Imines	imine	imino	$=NH$

[*]The carbon atom in parentheses in some groups belongs to an acyclic parent hydride. If there are no parentheses in the formula, the group as a whole is added to of the parent hydride.

Table 6.6 Characteristic groups always cited as prefixes

$-Br$	$-Cl$	$-F$	$-I$	$-NO_2$	$-NO$
bromo-	chloro-	fluoro-	iodo-	nitro-	nitroso-

Examples

6. 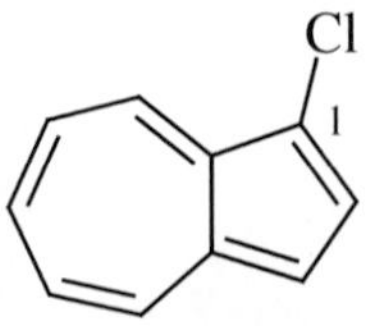 1-chloroazulene   7. 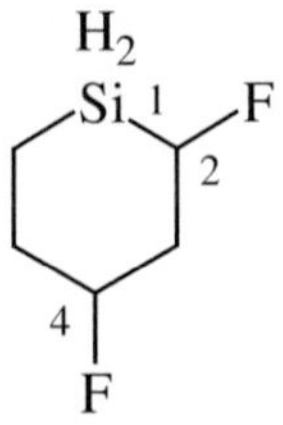 2,4-difluorosilinane   8. 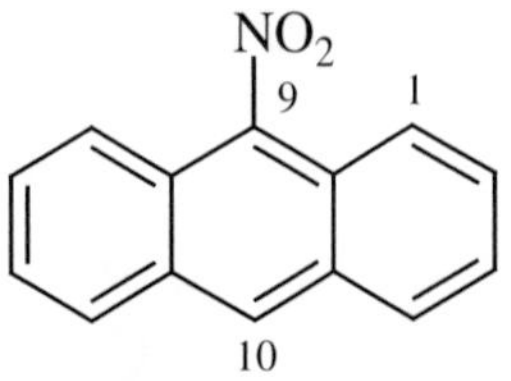 9-nitroanthracene

Normally, the names of all substituents are cited as prefixes at the front of the name of the parent hydride, but there are three exceptions. The two names toluene and styrene, are retained and can be used to name substituted derivatives, as long as the substitution is only on the ring and not expressed as a suffix (see Section 6.6.3).

Examples

9. 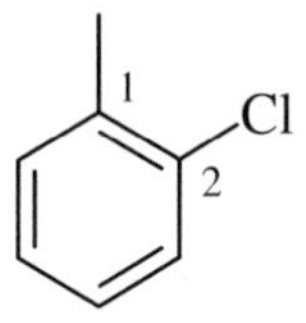 2-chlorotoluene   10. 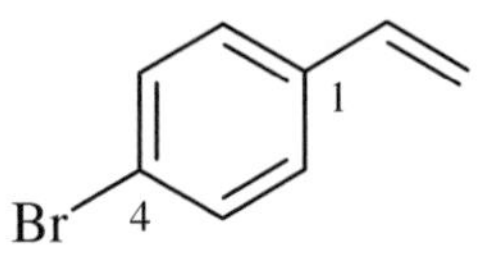 4-bromostyrene   11. 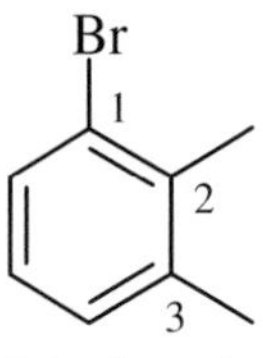 1-bromo-2,3-dimethylbenzene
(**not** 3-bromo-*o*-xylene)

75

Note that the name 3-bromo-*o*-xylene for Example 11 is not correct because the name xylene is not one of the cases of a parent hydride where an exceptional formulation is permitted. For benzene both types of formulation are allowed.

6.6.3   **Names of characteristic groups cited as suffixes and prefixes**

In the construction of names using substitutive nomenclature, suffixes must be used whenever possible for the principal groups. Prefixes are used to name all characteristic groups except principal groups. Lowest locants and multiplicative prefixes 'di', 'tri', 'tetra', *etc.*, are used following the general rules stated in Section 6.2. Suffixes and prefixes are listed in Tables 6.5 and 6.6. The names of some monofunctional compounds are shown below.

Examples

12. CH_3–CH_2–CH_2–OH	propan-1-ol
13. CH_3–CO–CH_2–CH_2–CH_3	pentan-2-one
14. CH_3–CH(OH)–CH(OH)–CH_3	butane-2,3-diol

The suffixes 'oic acid', 'amide', 'nitrile' and 'al' are used to name acyclic compounds having one or two characteristic groups. Locants are not necessary as these groups must be at the end of a chain. The suffixes 'carboxylic acid', 'carboxamide', 'carbonitrile' and 'carbaldehyde' are used if more than two groups are attached to a chain or if one or more groups are attached to a ring.

Examples

15. CH_3–CH_2–COOH   propanoic acid

16. OCH–CH_2–CH_2–CHO   butanedial (trivial name succinaldehyde)

17. CH_2(COOH)–CH(COOH)–CH_2(COOH)   propane-1,2,3-tricarboxylic acid (trivial name citric acid)

18. CH_3–CH_2–CH_2–CH_2–$CONH_2$ pentanamide

19. NC–$[CH_2]_4$–CN   hexanedinitrile (trivial name adiponitrile)

20. 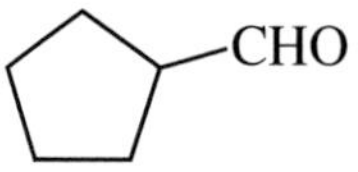 cyclopentanecarbaldehyde

21. 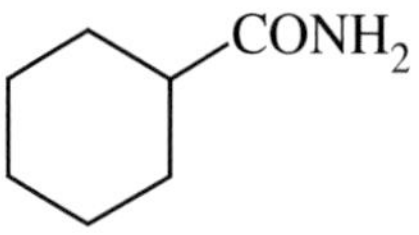 cyclohexanecarboxamide

22. 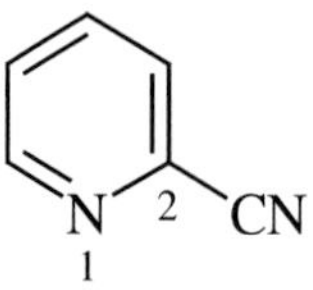 pyridine-2-carbonitrile

In the names of amines, the general use of suffixes and prefixes is not always observed. Normally, the suffix 'amine' would be added to the name of the parent hydride and engender names such as methanamine (CH_3–NH_2). Further substitution on the nitrogen atom would then be indicated by prefixes, leading to names that are explicit but cumbersome, such as N-methylmethanamine for CH_3–NH–CH_3 and N,N-dimethylmethanamine for CH_3–N(CH_3)$_2$, usually written (CH_3)$_3$N. The traditional names methylamine, dimethylamine and trimethylamine are much simpler. In these names, the term amine is not a suffix. It is, in fact, the name of the parent hydride NH_3 which now serves as the basis of substitutive names. Names such as diethylamine and triethylamine are representative of the preferred nomenclature. Diamines are named accordingly, for example, ethane-1,2-diamine for H_2N–CH_2–CH_2–NH_2 and propane-1,3-diamine for H_2N–[CH_2]$_3$–NH_2. There are allowed alternatives for these last two compounds: ethylenediamine and propane-1,3-diyldiamine.

Imines are named substitutively using the suffix 'imine'.

Example
    23. CH_3–CH=NH   ethanimine

The suffix 'one' is used to name ketones, and also related classes of the type X-CO-C, where X is a heteroatom such as O or N in a ring. For instance, the names of the lactones and the lactams, heterocyclic systems, are derived by adding the suffix 'one' to the name of the corresponding heterocycle. Specific suffixes 'olactone' and 'olactam' may also be used in simple cases.

Examples

24.

oxolan-2-one **or** butano-4-lactone

25.

azocan-2-one **or** heptano-7-lactam

To name polyfunctional compounds, two further rules are needed. First, since there can be only one type of characteristic group, which is cited as a suffix, that is, the principal group, the other groups must be cited as detachable prefixes. Second, the principal group is selected using the seniority order conveyed by Table 6.5, see Section 6.10.

6.7   NAMES OF FUNCTIONAL PARENT COMPOUNDS

Parent hydrides are alkanes, cycloalkanes and mancude ring systems having either trivial or systematic names. Some functional compounds, such as acetic acid, are still known under trivial names. These functional parent compound names are used in nomenclature like cyclic and acyclic parent hydride names, with one important difference. As they already contain characteristic groups prioritised to be cited as suffixes, **they can be further functionalised only by characteristic groups having lesser seniority**. These must then be cited as prefixes. There are very few such functional parent names recognised and retained, and some of them are given in the following list due to their importance in nomenclature.

Hydroxy compounds
C_6H_5–OH   phenol

Carbonyl compounds
CH_3–CO–CH_3 acetone

Carboxylic acids
H–COOH   formic acid
                              (for nomenclature purposes, the hydrogen atom linked to carbon is not regarded as substitutable)
CH_3–COOH   acetic acid
CH_2=CH–COOH   acrylic acid
HOOC–CH_2–COOH   malonic acid
HOOC–CH_2–CH_2–COOH   succinic acid
C_6H_5–COOH   benzoic acid
H_2N–COOH   carbamic acid
HOOC–COOH   oxalic acid
H_2N–CO–COOH   oxamic acid
1,2-C_6H_4(COOH)$_2$ phthalic acid
1,3-C_6H_4(COOH)$_2$ isophthalic acid
1,4-C_6H_4(COOH)$_2$ terephthalic acid

Amines
C_6H_5–NH_2 aniline

Acyclic polynitrogen compounds
H_2N–C(=NH)–NH_2 guanidine
H_2N–NH_2 hydrazine
H_2N–CO–NH_2 urea

The following compounds also have retained names.

Hydrocarbons
HC≡CH   acetylene
H_2C=C=CH_2 allene

Substituted benzenes

$C_6H_5-CH_3$	toluene
$C_6H_5-CH=CH_2$	styrene

Other names are **retained for unsubstituted compounds only** and compounds derived from them by substitution must be named systematically. The names are retained because of their wide use in biochemical and in polymer nomenclature. A few examples are given here.

Hydroxy compounds

$HO-CH_2-CH_2-OH$	ethylene glycol
$HO-CH_2-CH(OH)-CH_2-OH$	glycerol

Carboxylic acids

CH_3-CH_2-COOH	propionic acid
$CH_3-CH_2-CH_2-COOH$	butyric acid
$HOOC-[CH_2]_3-COOH$	glutaric acid
$HOOC-[CH_2]_4-COOH$	adipic acid
$H_2C=C(CH_3)-COOH$	methacrylic acid

Amines

$1,2-H_3C-C_6H_4-NH_2$	*o*-toluidine
$1,3-H_3C-C_6H_4-NH_2$	*m*-toluidine
$1,4-H_3C-C_6H_4-NH_2$	*p*-toluidine

The names propanoic acid (systematic) and propionic acid (retained) are both approved for the unsubstituted acid, CH_3-CH_2-COOH. Consequently, $Cl-CH_2-CH_2-COOH$ must be named systematically as 3-chloropropanoic acid. The compound $CH_3-CH(OH)-COOH$ is known as lactic acid; when it is substituted, for example in position 3 with a chlorine, its name is 3-chloro-2-hydroxypropanoic acid. Names such as 3-chloro-2-hydroxypropionic acid and 3-chlorolactic acid are not acceptable.

The retained names of carboxylic acids may also be modified to name amides, nitriles, and aldehydes, by changing the 'ic acid' ending to 'amide', 'onitrile' and 'aldehyde'. Names such as formamide, acetonitrile and propionaldehyde result. Of these names, only acetonitrile may be treated as a functional parent compound.

A further list of retained names, but including some of those discussed above, is given in Table 6.4.

6.8   NAMES OF RADICALS AND IONS

Radicals and ions are not formed by a substitution operation, but by subtraction or addition of hydrogen atoms, hydrons or hydride ions. Their names are formed using appropriate suffixes and prefixes, many of which are listed in Table 6.7.

Examples

1. $CH_3\cdot$	methyl
2. $CH_2^{2\cdot}$	methylidene **or** carbene
3. $\cdot CH_2-CH_2\cdot$	ethane-1,2-diyl

4. $CH_3-CH_2^-$ ethanide **or** ethyl anion

5. CH_5^+ methanium

6. CH_3^+ methylium **or** methyl cation

These suffixes are cumulative, meaning that more than one can be present in a name, for instance, to represent radical ions or derivatives of ionic compounds. Radicals come before anions, which in turn come before cations, in the order of seniority for citation as suffixes (see Table 6.7). Suffixes for radicals are cited last in names.

Examples

7. $CH_2^{\bullet+}$ methyliumyl

8. $CH_4^{\bullet+}$ methaniumyl

9. $^+CH_4-CH_2^{\bullet}$ ethan-2-ium-1-yl

10. $^-CH_2-CH_2^{\bullet}$ ethan-2-id-1-yl

11. $CH_3-NH_2^{\bullet+}$ methanamin-*N*-ium-*N*-yl

The terms 'radical cation' or 'radical ion(+)' may be added to the name of a parent hydride when positions of free valences and/or charges are not known, or when it is not desirable to specify them.

Example

12. $(C_2H_6)^{\bullet+}$ ethane radical cation **or** ethane radical ion(1+)

Salts of acids and alcohols are described using binary names composed of an anion name and a cation name (see Section 5.2). The name of the anion is obtained by using the suffix 'ate', just as used in obtaining the names of anions derived from 'ic acids' (see Section 5.4.) but here operating on the systematic name of the alcohol. Salts of amines are also named using a binary procedure, employing the suffix 'ium' and the locant *N*, as described above.

Examples

13. $CH_3COO^-\,Na^+$ sodium acetate

14. $CH_3O^-\,K^+$ potassium methanolate

15. $C_6H_5NH_3^+\,Cl^-$ anilin-*N*-ium chloride

This completes the discussion of the general derivation of substitutive names. The content is relatively comprehensive, and for many students and teachers contains too much detail. Care should be taken in selecting teaching material appropriate to the needs of the potential users. Section 6.10 provides a summary of the methodology outlined above.

Table 6.7 Some suffixes used to name radicals and ions

	Operation	Suffix
Radicals	Loss of $H^{\bullet}$	yl
	Loss of $2H^{\bullet}$	ylidene or diyl*
Anions	Loss of H^+	ide
		(salts and esters generally receive the suffix ate)
Cations	Loss of H^-	ylium
	Addition of H^+	ium

*The suffix 'ylidene' is used to represent radicals such as $R-CH^{2\bullet}$; the suffix 'diyl' is used when the two hydrogen atoms have been removed from two different skeleton atoms.

6.9   STEREOCHEMICAL DESCRIPTORS

The use of stereochemical descriptors in both names and formulae is described in detail in Chapter 3.

6.10   SUMMARY OF THE GENERAL PROCEDURE TO NAME COMPOUNDS USING SUBSTITUTIVE NOMENCLATURE

To name a compound using substitutive nomenclature, the following sequence of operations is applied.

1. Confirm that substitutive nomenclature is to be used.

2. Identify any principal characteristic group of the compound concerned, so that this can be cited in the name as a suffix.

3. Determine the identity of the parent hydride or functional parent compound. Its name must include any necessary non-detachable prefixes.

4. Name parent hydride and suffix or the functional parent compound, and number it as far as possible.

5. Determine structural features to be cited as prefixes.

6. Name the substituents to be cited as detachable prefixes, identify the required 'hydro' prefixes and 'ene' and 'yne' endings, and complete the numbering.

7. Assemble the components in a complete name using alphabetical order for all detachable prefixes.

8. Determine the configuration and add appropriate stereodescriptors.

Examples

1. CH_3–CO–CH_2–CH(OH)–CH_3 4-hydroxypentan-2-one

2. H_2N–CH_2–CH_2–OH   2-aminoethan-1-ol

3. OHC–CH(CH_3)–CH_2–CH_2–COOH   4-methyl-5-oxopentanoic acid

4. H_2N–CO–CH_2–CH_2–COOH   4-amino-4-oxobutanoic acid

5. NC–CH_2–CH=CH–$CONH_2$ 4-cyanobut-2-enamide

6.   4-formylcyclohexane-1-carboxylic acid

7.   2-carbamoylcyclopentane-1-carboxylic acid

8.   (3R)-6-amino-3-fluoro-6-methyl▶ heptanoic acid

Note the use of the symbol ▶ in Example 8. In this text, and in this text alone, this symbol is used to show a line-break imposed solely by demands of typesetting. It is not a nomenclature symbol and should never be reproduced as part of this or any other name. See also p. 87.

Both suffixes and prefixes are necessary to name structures composed of various sub-units, for instance when characteristic groups are situated on side-chains. The principal chain is chosen in accordance with the selection criteria in Table 6.1 and is the chain supporting the greatest number of principle characteristic groups. If this is not decisive, the longest chain, criterion (a), is chosen. Further criteria (b), (c), *etc.*, are used as required.

Example

9.

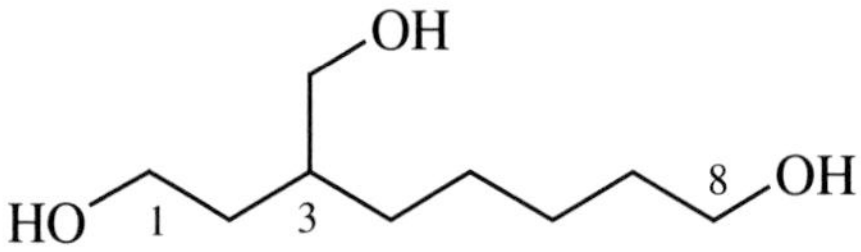

3-(hydroxymethyl)octane-1,8-diol

When a choice of lowest locants has to be made, these are assigned first to heteroatoms in rings, then to the principal characterstic group, then to positions of unsaturation, and finally to substituents cited as detachable prefixes.

Examples

10.

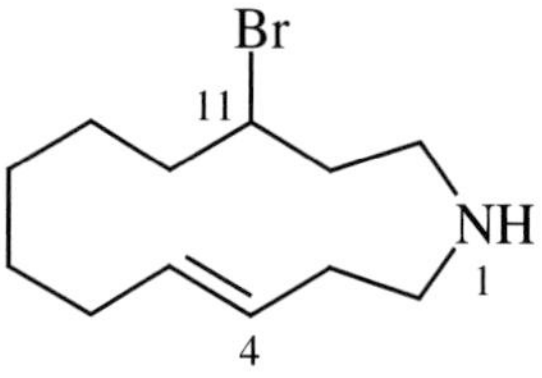

11-bromo-1-azacyclotridec-4-ene

11.

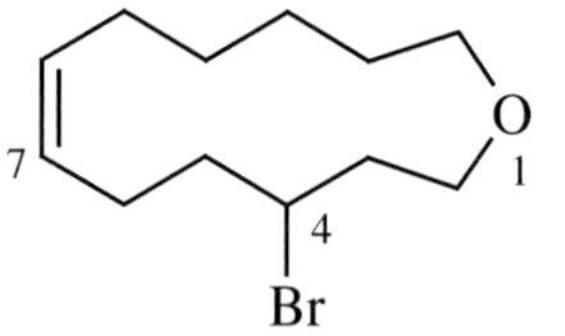

4-bromo-1-oxacyclotridec-7-ene

12.

$$HC{\equiv}C{-}CH_2{-}\underset{\underset{NH_2}{|}}{CH}{-}CH_2{-}CH_2{-}CH_2{-}NO_2$$

7-nitrohept-1-yn-4-amine

13.

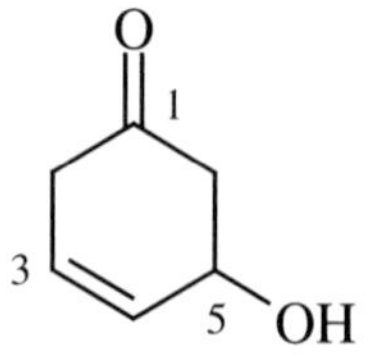

5-hydroxycyclohex-3-en-1-one

6.11 FUNCTIONAL CLASS NOMENCLATURE

Substitutive nomenclature is the nomenclature of choice in organic chemistry. However it cannot be used to name all classes of compound. Salts, esters, acid halides and anhydrides cannot be named substitutively when the characteristic group is chosen as the principal group and **functional class names**, which were formerly called **radicofunctional names**, are used. Functional class nomenclature is a binary system widely used in organic chemistry, as mentioned in Chapter 4. Binary names are composed of two parts. The names of salts of carboxylic acids are binary, as in the name of the salt, sodium acetate. Names of esters and acyl halides are constructed in a similar way: methyl acetate for $CH_3-COOCH_3$, methyl chloroacetate for $ClCH_2-COOCH_3$, acetyl chloride for $CH_3-CO-Cl$, and benzoyl bromide for $C_6H_5-CO-Br$. These groups are cited as prefixes when a group is present which is of higher priority for citation as principal group.

Examples

1. $ClCO-CH_2-CH_2-COOH$ 4-chloro-4-oxobutanoic acid
2. $C_6H_5-COO-[CH_2]_3-COOH$ 4-(benzoyloxy)butanoic acid

To name anhydrides using binary nomenclature functional modifiers are employed, *e.g.*, acetic anhydride for $CH_3-CO-O-CO-CH_3$.

6.12 NAME INTERPRETATION

Organic nomenclature can be a complex and intimidating subject. Nevertheless, it is pointless for an author to apply it if the reader cannot decipher a name or a formula in order to determine the structure of the compound under discussion. Chapter 13 provides a general guide as to how to do this.

In general, the reader must determine the type of nomenclature (substitutive or functional class) being used, and identify a parent hydride name that has been modified by any of a large number of affixes (prefixes, infixes, and suffixes) of various kinds, so it is necessary to be familiar with these basic nomenclature entities. A systematic approach to the decipherment is based upon recognising these entities in a name, and assessing their significance. This is described in Chapter 13, but it is also important to recognise that nomenclature terms and conventions do develop and change. Some familiarity with older nomenclature may on occasion be of use, especially when reading earlier literature.

REFERENCES

1. The Lambda Convention (IUPAC Recommendations 1983), *Pure and Applied Chemistry*, 1984, **56**, 769–778.
2. Nomenclature of Fused and Bridged Fused Ring Systems (IUPAC Recommendations 1998), *Pure and Applied Chemistry*, 1998, **70**(1), 143–216.

For further and new information on organic nomenclature consult http://www.chem.qmul.ac.uk/iupac/

7 Additive Nomenclature and the Nomenclature of Coordination Compounds

7.1 INTRODUCTION

In the nineteenth and early twentieth centuries, numerous compounds were prepared that were formally a combination of metal salts with other stable ions or molecules. For example, compounds were prepared that contained cobalt(III) chloride and also three, or four, or five, or even six molecules of ammonia per cobalt ion. This kind of compound became known as a complex salt, because it was recognized that the structure was likely to be more complicated than that of more conventional salts. In some cases more than one compound with the same composition could be isolated, so that isomers existed. Analysis of the possibilities for isomerism of this kind was a key in the development of inorganic chemistry, and the work of Alfred Werner in particular led to coordination theory and to the award of the Nobel Prize in 1913.

According to this theory, mononuclear coordination compounds are considered to consist of a central atom, often a metal ion, which is bonded to surrounding small molecules or ions, called ligands. Structural diagrams are used to represent these molecules/ions and these show the bonds to the central atom. The names and formulae of coordination compounds are constructed using the names and formulae of the central atoms and of the bound molecules or ions. The name of the whole entity is put together by adding the ligand names to that of the central atom. The kind of nomenclature used to produce names in this way is known as additive nomenclature. The structures proposed for coordination compounds are reflected in the nomenclature that was developed in order to name them.

The concept of **primary valence**, or **Hauptvalenz** in the original German, was used by Werner to account for the formal charge on the central atom/ion and therefore also for the number of counterbalancing anions in compounds, whereas **secondary valence** or **Nebenvalenz** was used to describe the numbers of other molecules/ions that form part of the coordination compound. In modern terminology for coordination compounds, the primary valence represents the **oxidation state** of the central atom, while the secondary valence of a central atom is normally related to the number of bonds formed by the central atom to the surrounding atoms or ions – the

Principles of Chemical Nomenclature: A Guide to IUPAC Recommendations
Edited by G. J. Leigh
© International Union of Pure and Applied Chemistry 2011
Published by the Royal Society of Chemistry, www.rsc.org

coordination number. The terms 'primary valence' and 'secondary valence' are no longer recommended.

In this Chapter we describe the conventions that are used commonly to represent coordination compounds and to provide appropriate rules and guidelines for use of each style. The discussion is limited to compounds with only one kind of central atom in each coordination entity, which means that there is either only one central atom or, if there are more, that they are in identical coordination environments. The presence of different kinds of central atom within the same coordination entity leads to significant complications in the formation of the name (see the *Nomenclature of Inorganic Chemistry, IUPAC Recommendations 2005*, particularly Chapter 9, for more detail on how such situations are treated). The flow charts and rules presented here will guide the user in the construction of formulae and names from structural diagrams, initially for straight-forward cases and then for more complicated examples. The numerous examples demonstrate sound practice in all these aspects.

Examples 1 and 2 show a number of different ways in which coordination compounds can be described. In Example 1, which can be variously represented by the formula of an addition compound, $CoCl_3 \cdot 6NH_3$, by the formula of a complex compound, $[Co(NH_3)_6]Cl_3$ or by an additive name, hexaamminecobalt(III) chloride, the formal charge on the cobalt is $3+$, so that the oxidation state is III. Oxidation states are conventionally designated by roman numerals, and are always assumed to be positive unless specifically denoted as negative, as with a chloride ion, oxidation state $-I$. The coordination number of the cobalt is 6 (there are six ammonia ligands coordinated to the central cobalt atom), and the geometry around the cobalt ion is octahedral. In Example 2, described in old literature as $PtCl_2 \cdot 2KCl$, and more recently as $K_2[PtCl_4]$ or potassium tetrachloridoplatinate(II), the oxidation state of the platinum ion is II, the coordination number is 4, and the geometry around the central atom is square planar. Tetrachloroplatinate is the hitherto commonly used name for the anion, though according to the current IUPAC recommendations it is now more properly called tetrachloridoplatinate(II).

Examples

1. A compound with a complex cation

Name: (*OC*-6)-hexaamminecobalt(III) trichloride
Formula: $[Co(NH_3)_6]Cl_3$

2. A compound with a complex anion

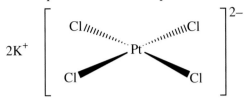

Name: dipotassium (*SP*-4)-tetrachloridoplatinate
Formula: $K_2[PtCl_4]$

The devices (*OC*-6) and (*SP*-4) describe the configurations of the complexes, as detailed in Sections 3.9 and 7.5.7. The charges shown in Examples 1 and 2 are sometimes omitted from the structural formula if they are reckoned to be self-evident. In formulae, the coordination entity is always enclosed by square brackets.

These examples illustrate the most commonly used representations for coordination compounds: structural diagrams, stoichiometric formulae, molecular formulae, and additive names. The different representations provide different amounts of information about the structure. Structural diagrams provide detailed information about a coordination compound, showing which atoms are connected by bonds and how those atoms are located in space relative to each other. However, such diagrams are not always easily incorporated into text. Names can also provide this level of information if all the appropriate conventions are employed, but they may be rather long and inconvenient to use. Formulae are often more convenient, but provide somewhat less structural information. The amount of detail provided in any given case should depend on the need for adequate comprehension.

WARNING: many names and formulae are very long, and cannot always be comfortably accommodated on a single line, requiring a line-break. In normal text this would be indicated by a hyphen, but a hyphen is a symbol in nomenclature. **In this text alone** we use the symbol ▶ to indicate a line break in names or formulae where in normal text a hyphen would be used. This symbol is not part of the name or formula, and should be omitted when reproducing parts of this text and should never be used when constructing a name or formula.

7.2   DEFINITIONS

7.2.1   Coordination compounds, complexes and coordination entities

A **coordination compound** is a compound that contains a central atom (Section 7.2.2) and surrounding ligands (Section 7.2.3). This part of a coordination compound is known as a **complex** or a **coordination entity**. The cations of Section 7.1, Example 1, and of Section 7.2.1, Example 2, and the anions of Section 7.1. Example 2, and Section 7.2.1, Example 2, as well as the neutral species in Section 7.2.1, Example 1, are all coordination complexes.

If any of the ligands other than CN^-, CO, CS have a carbon atom attached to the central atom, then the compound may also be considered to be an organometallic compound (see Chapter 8) as well as a coordination compound and organometallic compounds may have their own specialised nomenclature. Example 3 represents such a compound.

Examples

1. A neutral complex

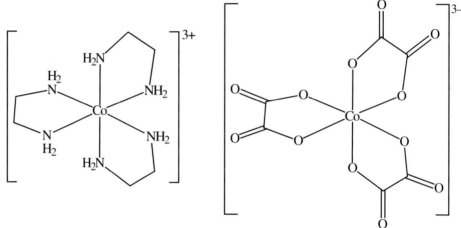

Name: (*SPY*-5-23′)-bis(aminoethanoato)aquacopper(II) **or**
(*SPY*-5-23′)-bis(aminoacetato)aquacopper(II)
Formula: [Cu(gly)₂(OH₂)]

2. A compound where both the anion and cation are complex ions

Name: (*OC*-6)-tris(ethane-1,2-diamine)cobalt(III) (*OC*-6)-tris(ethanedioato)cobaltate(III)
Formula: [Co(en)₃][Co(ox)₃]

3. An organometallic compound

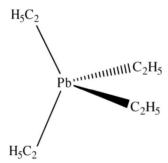

Name: tetraethanidolead(IV)
Formula: [Pb(C₂H₅)₄]

7.2.2   Central atoms

The **central atom** is the atom or ion around which the rest of the coordination entity or complex is built. It is often a metal ion, it is generally close to the physical centre of the structure, and it is often the atom/ion with the highest coordination number. The central atom in Section 7.2.1, Example 1, is copper(II).

In Example 2 cobalt(III) is the central atom for both the cation and the anion. A coordination entity may have more than one central atom, but here the discussion is limited to those simpler cases where there is just one kind of central atom in each coordination entity. However, for complicated structures a suitable name is often easier to form if more central atoms are treated as coordination centres.

7.2.3 Ligands and donor atoms

A **ligand** is an atom, molecule or ion that is bonded to a central atom by a covalent bond. Often the electrons that are shared between the ligand and the central atom can be considered formally to have been provided by the ligand, but this need not actually be the case. A ligand may be connected to a central atom through more than one atom and it may be connected to more than one central atom. The **donor atoms** (or ligating atoms) are atoms in ligands that are directly attached to the central atom.

7.2.4 Chelation and denticity

A **chelating ligand** is one which binds to the central atom through more than one (non-contiguous) donor atom. The **denticity** of a ligand is the number of different donor atoms that are bound to the central atom. Ligands in which con-tiguous (*i.e.*, sequences of neighbouring) donor atoms are bound to a central atom, as are common in transition-metal organometallic compounds (Chapter 8) are not generally regarded as chelating ligands. With the exception of the water ligand in Example 1, all the ligands in Section 7.2.1, Examples 1 and 2, are chelating ligands, but those in Example 3 are not.

7.2.5 Bridging ligands

A **bridging ligand** is one that is bound to more than one central atom, forming a bridge between them. The central atoms may be bound to the same donor atom or to different donor atoms within the bridging ligand. Examples 4, 5 and 6 below illustrate both cases. The designator μ (Greek mu) is used to indicate bridging ligands.

Examples
    4. A complex compound containing bridging chloride ions

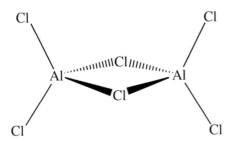

Name: di-μ-chlorido-bis[dichloridoaluminium(III)]
Formula: $[Al_2Cl_6]$ **or** $[Cl_2Al(\mu\text{-}Cl)_2AlCl_2]$

5. A complex anion containing bridging sulfide ions

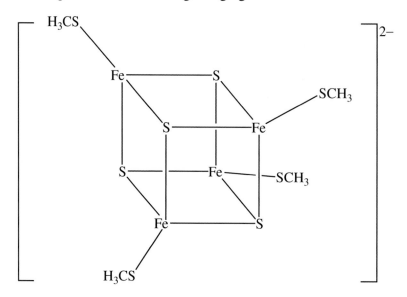

Name: *tetrahedro*-tetra-μ_3-sulfido-tetrakis(methanethiolatoferrate)(2−)
Formula: $[Fe_4(\mu_3\text{-}S)_4(SCH_3)_4]^{2-}$

6. A complex in which donor atoms of the bridging ligand are coordinating to the different central atoms

Name: μ-ethane-1,2-diamine-1κN^1:2κN^2-bis[pentaamminecobalt(III)]
Formula: $[(NH_3)_5Co(\mu\text{-}NH_2CH_2CH_2NH_2)Co(NH_3)_5]^{6+}$

7.2.6   Metal-metal bonding

Numerous compounds exist in which central atoms are bonded directly to each other without any bridging ligand. From electron-counting or bond-length considerations, such bonds may be considered to be single, double, triple, and quadruple metal-metal bonds, but in nomenclature no attempt is made to indicate the multiplicity of such bonds.

Examples

7. A compound in which the metal atoms are considered to be joined to each other by a quadruple bond

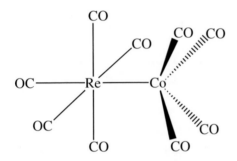

Name: dicaesium bis(tetrachloridorhenate)(*Re—Re*)
Formula: $Cs_2[Cl_4ReReCl_4]$ or $Cs_2[Re_2Cl_8]$

8. A compound in which different metal atoms are joined by a single bond

Name: nonacarbonyl-1$\kappa^5 C$,2$\kappa^4 C$-rheniumcobalt(*Re—Co*)
Formula: $[(OC)_5ReCo(CO)_4]$

7.2.7 Relative and absolute configurations

For coordination compounds, the relative and absolute configurations describe the spatial arrangement of atoms around the central atom. This is done by first specifying the ideal geometry closest to that of the observed arrangement of donor atoms and then the coordination number. These are indicated by using a **polyhedral symbol** (see Section 7.5.7 and Table P5). The relative positions of donor atoms around the central atom can be described with a **configuration index** (Section 3.9), though the absolute configurations of the donor atoms may also be described (Section 3.9).

7.3 STRUCTURAL DIAGRAMS OF COORDINATION COMPOUNDS

A good structural diagram should indicate clearly which atoms are bonded to which, and may even indicate the bond order of such bonds, provided they are reckoned to have integral values, single, double, triple, or quadruple. Bonds with non-integral bond orders are not routinely differentiated from other bonds. Structural diagrams can also be used to represent the relative configurations and absolute configurations of a structure, so that they can be used to distinguish between diastereoisomers and between enantiomers (see also Section 3.9).

The general guidelines for drawing structures of chemical compounds apply to coordination compounds. Structural diagrams should be clear and unambiguous. A complete account of current IUPAC recommendations on this subject is to be found in reference 1. The choices of bond representation, font size for atom labels, and appropriate positioning of bonds and atom labels are particularly important. For coordination compounds, particular care must be taken when describing the geometry around the central atom. The higher coordination numbers can make it difficult to avoid crowding and ambiguity in drawings. The structural diagrams in this Chapter are examples of appropriate representations of coordination compounds.

7.4   FORMULAE OF COORDINATION COMPOUNDS

7.4.1   Introduction

Formulae are short-hand forms of the more explicit structural formulae. One goal of a formula is to provide an abbreviated representation of a compound. Subtleties of coordination mode are more usually presented in structural diagrams or in names. Formulae are shorter than full names, which may become rather unwieldy for coordination compounds. The abbreviated nature of formulae means that sometimes they are not able to convey as much information about a chemical structure as a full name or structural formula. Examples 1–8 of Table 7.2a illustrate the process of formula construction.

Formulae are made up of element symbols for central atoms and simple ligands, but more complicated, polyatomic ligands are often represented by abbreviations. For example, ethane-1,2-diamine, ethanedioate (or oxalate), and aminoethanoate (or glycinate) are often abbreviated within formulae as en, ox and gly, respectively (see Section 7.2.1, Examples 1 and 2). A list of commonly accepted abbreviations is presented in Table 7.1 (see page 93) and a longer list may be found in the *Nomenclature of Inorganic Chemistry, IUPAC Recommendations 2005*. Similarly, in organometallic nomenclature a cyclopentadienyl ligand is often represented by the abbreviation Cp (see Chapter 8). For convenience, many authors will create abbreviations for use when representing complicated ligands within formulae, but care should be taken to avoid use of abbreviations already established for other ligands. In any but the most widely accepted cases, such as those noted here in Table 7.1, abbreviations should be clearly defined upon the first time of use in a given text. An abbreviation for a ligand name should be based on key letters of the full name.

The procedure for producing a formula from a structural diagram is very similar to that used to create a name, and Figure 7.1 shows a flow chart that should be used to aid formula construction. Detailed examples are laid out in Table 7.2a, Examples 1–8. The treatment here is limited to situations where there is only one kind of central atom within each coordination entity. It is less common to define the coordination geometry in formulae, but if this is required it is done in exactly the same way as for names (see Section 7.5.7 below and the related Examples).

Table 7.1 Selected ligand abbreviations for use in formulae

Abbreviation	Systematic ligand name	Other name from which abbreviation was originally derived[a]
acac	2,4-dioxopentan-3-ido	acetylacetonato
ade	9*H*-purin-6-amine	adenine
aet	2-aminoethane-1-thiolato	
ala	2-aminopropanoato	alaninato
amp	adenosine 5′-phosphato(2−)	adenosine monophosphato
[9]aneN₃ (also tacn)	1,4,7-triazonane	1,4,7-triazacyclononane
[12]aneN₄ (also cyclen)	1,4,7,10-tetraazacyclododecane	
[14]aneN₄ (also cyclam)	1,4,8,11-tetraazacyclotetradecane	
[9]aneS₃	1,4,7-trithionane	1,4,7-trithiacyclononane
[12]aneS₄	1,4,7,10-tetrathiacyclododecane	
arg	2-amino-5-guanidinopentanoato	argininato
asn	2,4-diamino-4-oxobutanoato	asparaginato
asp	2-aminobutanedioato	aspartato
atp	adenosine 5′-triphosphato(4−)	
binap	1,1′-binaphthalene-2,2′-diylbis(diphenylphosphane)	
bpy	2,2′-bipyridine	
4,4′-bpy	4,4′-bipyridine	
Bu	butyl	
bzim	1*H*-benzimidazol-1-ido	
Bz[b]	benzoyl	
cat	benzene-1,2-bis(olato)	catecholato
chxn (also dach)	cyclohexane-1,2-diamine	
C₅Me₅[c]	pentamethyl-η⁵-cyclopentadienyl	
cod	cycloocta-1,5-diene	
cot	cycloocta-1,3,5,7-tetraene	
Cp	η⁵-cyclopentadienyl	
18-crown-6	1,4,7,10,13,16-hexaoxacyclooctadecane	
crypt-211	4,7,13,18-tetraoxa-1,10-diazabicyclo[8.5.5]icosane	cryptand 211
crypt-222	4,7,13,16,21,24-hexaoxa-1,10-diazabicyclo▶ [8.8.8]hexacosane	cryptand 222
Cy	cyclohexyl	
cyclam (see [14]aneN₄)	1,4,8,11-tetraazacyclotetradecane	
cyclen (see [12]aneN₄)	1,4,7,10-tetraazacyclododecane	
cyt	4-aminopyrimidin-2(1*H*)-one	cytosine
dabco	1,4-diazabicyclo[2.2.2]octane	
dach (see chxn)		
dbm	1,3-dioxo-1,3-diphenylpropan-2-ido	dibenzoylmethanato
depe	ethane-1,2-diylbis(diethylphosphane)	1,2-bis(diethylphosphino)ethane
dien	*N*-(2-aminoethyl)ethane-1,2-diamine	diethylenetriamine
diox	1,4-dioxane	
dma	*N,N*-dimethylacetamide	dimethylacetamide
dme	1,2-dimethoxyethane	
dmf	*N,N*-dimethylformamide	
dmg	butane-2,3-diylidenebis(azanolato)	dimethylglyoximato
dmpe	ethane-1,2-diylbis(dimethylphosphane)	1,2-bis(dimethylphosphino)ethane
dmso	(methylsulfinyl)methane	dimethyl sulfoxide
dpm	2,2,6,6-tetramethyl-3,5-dioxoheptan-4-ido	dipivaloylmethanato
dppe	ethane-1,2-diylbis(diphenylphosphane)	1,2-bis(diphenylphosphino)ethane
dtpa	2,2′,2″,2‴-[(carboxylatomethyl)azanediylbis(ethane-▶ 2,1-diylnitrilo)]tetraacetato	diethylenetriaminepentaacetato
ea	2-aminoethan-1-olato	ethanolaminato
edda	2,2′-[ethane-1,2-diylbis(azanediyl)]diacetato	ethylenediaminediacetato
edta	2,2′,2″,2‴-(ethane-1,2-diyldinitrilo)tetraacetato	ethylenediaminetetraacetato
en	ethane-1,2-diamine	
Et	ethyl	
Et₂dtc	*N,N*-diethylcarbamodithioato	*N,N*-diethyldithiocarbamato
gln	2,5-diamino-5-oxopentanoato	glutaminato
glu	2-aminopentanedioato	glutamato
gly	aminoethanoato, **or** 2-aminoacetato	glycinato

CHAPTER 7

Table 7.1 (*Continued*)

Abbreviation	Systematic ligand name	Other name from which abbreviation was originally derived[a]
gua	2-amino-9*H*-purin-6(1*H*)-one	guanine
guo	2-amino-9-β-D-ribofuranosyl-9*H*-purin-6(1*H*)-one	guanosine
hfa	1,1,1,5,5,5-hexafluoropentane-2,4-dioxopentan-3-ido	hexafluoroacetylacetonato
his	2-amino-3-(imidazol-4-yl)propanoato	histidinato
hmpa	*N,N,N′,N′,N″,N″*-hexamethylphosphoric triamide	hexamethylphosphoramide
ida	2,2′-azanediyldiacetato	iminodiacetato
ile	2-amino-3-methylpentanoato (all four possible stereoisomers)	isoleucinato **or** alloisoleucinato
im	imidazol-1-ido	
isn	pyridine-4-carboxamide	isonicotinamide
leu	2-amino-4-methylpentanoato	leucinato
lut	2,6-dimethylpyridine	lutidine
lys	2,6-diaminohexanoato	lysinato
Me	methyl **or** methanido	
2-Mepy	2-methylpyridine	
met	2-amino-4-(methylsulfanyl)butanoato	methioninato
nbd	bicyclo[2.2.1]hepta-2,5-diene	norbornadiene
nia	pyridine-3-carboxamide	nicotinamide
nta	2,2′,2″-nitrilotriacetato	
oep	2,3,7,8,12,13,17,18-octaethylporphyrin-21,23-diido	
ox	ethanedioato	oxalato
pc	phthalocyanine-29,31-diido	
Ph	phenyl	
phe	2-amino-3-phenylpropanoato	phenylalaninato
phen	1,10-phenanthroline	
pip	piperidine	
pn	propane-1,2-diamine	
ppIX	2,18-bis(2-carboxyethyl)-3,7,12,17-tetramethyl-▶ 8,13-divinylporphyrin-21,23-diido	protoporphyrinato IX
pro	pyrrolidine-2-carboxylato	prolinato
ptn	pentane-2,4-diamine	
py	pyridine	
pyz	pyrazine	
pz	1*H*-pyrazol-1-ido	
quin	quinolin-8-olato	
sal	2-hydroxybenzoato	salicylato
salen	2,2′-[(ethane-1,2-diyl)bis▶ (azanylylidenemethanylylidene)]diphenolato	bis(salicylidene)ethylenediaminato
salgly	*N*-(2-oxidobenzylidene)glycinato	salicylideneglycinato
sep[d]	1,3,6,8,10,13,16,19-octaazabicyclo[6.6.6]icosane	
ser	2-amino-3-hydroxypropanoato	serinato
tacn (see [9]aneN₃)		
tap	propane-1,2,3-triamine	1,2,3-triaminopropane
tcne	ethenetetracarbonitrile	tetracyanoethylene
tcnq	2,2′-(cyclohexa-2,5-diene-1,4-diylidene)▶ dipropanedinitrile	tetracyanoquinodimethane
terpy	2,2′:6′,2″-terpyridine	terpyridine
2,3,2-tet	*N,N′*-bis(2-aminoethyl)propane-1,3-diamine	1,4,8,11-tetraazaundecane
3,3,3-tet	*N,N′*-bis(3-aminopropyl)propane-1,3-diamine	1,5,9,13-tetraazatridecane
tetren	*N,N′*-(azanediyldiethane-2,1-diyl)di(ethane-1,2-diamine)	tetraethylenepentamine
tfa	trifluoroacetato	
thf	oxolane	tetrahydrofuran
thr	2-amino-3-hydroxybutanoato	threoninato
thy	5-methylpyrimidine-2,4(1*H*,3*H*)-dione	thymine
tmen	*N,N,N′,N′*-tetramethylethane-1,2-diamine	
tmp	5,10,15,20-tetrakis(2,4,6-trimethylphenyl)porphyrin-▶ 21,23-diido	5,10,15,20-tetramesitylporphyrin-▶ 21,23-diido
tn	propane-1,3-diamine	trimethylenediamine

Table 7.1 (*Continued*)

Abbreviation	Systematic ligand name	Other name from which abbreviation was originally derived[a]
(*o*-, *m*- or *p*-)tol	2-, 3- or 4-methylphenyl	(*o*-, *m*- or *p*-) tolyl
Tp	hydridotris(pyrazolido-*N*)borato(1–), **or** tris(1*H*-pyrazol-1-yl)boranuido	hydrotris(pyrazolyl)borato
Tp′ [e]	hydridotris(3,5-dimethylpyrazolido-*N*)borato(1–)	hydrotris(3,5-▶ dimethylpyrazolyl)borato
tpp	5,10,15,20-tetraphenylporphyrin-21,23-diido	
tren	*N*,*N*-bis(2-aminoethyl)ethane-1,2-diamine	tris(2-aminoethyl)amine
trien	*N*,*N*′-bis(2-aminoethyl)ethane-1,2-diamine	triethylenetetramine
trp	2-amino-3-(1*H*-indol-3-yl)propanoato	tryptophanato
tu	thiourea	
tyr	2-amino-3-(4-hydroxyphenyl)propanoato	tyrosinato
ura	pyrimidine-2,4(1*H*,3*H*)-dione	uracil
val	2-amino-3-methylbutanoato	valinato

[a]Many of these names upon which abbreviations were originally based are no longer recommended.

[b]The abbreviation Bz has often been used for benzyl, for which IUPAC has recommended Bzl and Bn. Use of the unambiguous alternatives, PhCO and PhCH$_2$, is therefore preferable.

[c]The use of the abbreviation Cp* for pentamethylcyclopentadienyl is discouraged. It can lead to confusion because the asterisk, *, is also used to represent an excited state, an optically active substance, *etc.*

[d]The abbreviation derives from the non-systematic name sepulchrate which incorrectly implies that the ligand is anionic.

[e]The use of Tp′ is preferred to Tp* for the reasons given in footnote c. A general procedure for abbreviating substituted hydridotris(pyrazolido-*N*)borate ligands has been proposed (see S. Trofimenko, *Chemical Reviews*, 1993, **93**(3), 943–980). For example, Tp′ becomes Tp^{Me_2}, the superscript denoting the methyl groups at the 3- and 5-positions of the pyrazole rings.

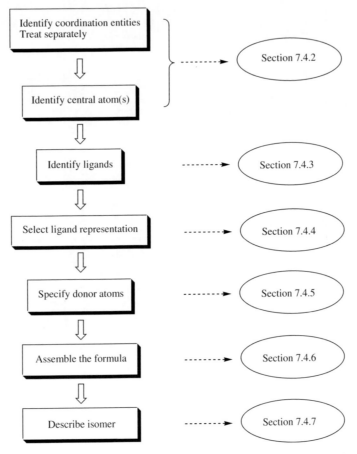

Figure 7.1 A flow chart showing the steps in producing a formula from the structure of a coordination compound.

Table 7.2a Examples of constructing linear formulae from structural formulae. (As discussed in Section 7.1, in this text only, the symbol ▶ merely indicates a line break and is **NOT** a nomenclature symbol.)

Naming step	Example 1	Example 2	Example 3	Example 4
Structure for which formula required				
Identify central atom(s)	Co	Pt	Co	S
Identify and name ligands	NH_3 OH_2	Cl triphenylphosphane → PPh_3	ethane-1,2-▶diyldinitrilotetraacetate → edta	O O_2
Specify donor atom(s)	*not required for ammonia and water*	*not required for chloride or triphenylphosphane*	*not required for edta unless not bound in a hexadentate fashion*	*not required for oxide* μ-O_2
Assemble the formula	$[Co(NH_3)_5(OH_2)]Cl_3$	$[PtCl_2(PPh_3)_2]$	$Ba[Co(edta)]_2$	$[O_3S(\mu\text{-}O_2)SO_3]^{2-}$
Describe the coordination geometry	*not required*	*trans-*	*not required*	*not required*

Naming step	Example 5	Example 6	Example 7	Example 8
Structure of which formula required				
Identify central atom(s)	Co	$2 \times$ Re	Mo	$4 \times$ Fe
Identify and name ligands	$H_2NC(CH_3)_2C(CH_3)_2NH_2$ O_2	Cl	Cl 1,4,8,12-tetra▶ thiacyclopentadecane → [15]aneS$_4$	S SCH$_3$
Specify donor atom(s)	not usually done for amines – all N atoms assumed to coordinate η^2-O_2	not required for Cl	not required for Cl [15]aneS$_4$-$\kappa^3 S^1,S^4,S^8$	μ_3-S not required for SCH$_3$
Assemble the formula	$[Co\{H_2NC(CH_3)_2C(CH_3)_2NH_2\}_2(\eta^2\text{-}O_2)]^+$ ▶	$Cs_2[Cl_4ReReCl_4]$	$[Mo([15]aneS_4\text{-}\kappa^3 S^1,S^4,S^8)Cl_3]$	$[Fe_4(\mu_3\text{-}S)_4(SCH_3)_4]^{2-}$
Indicate the coordination geometry	not required	not required	not required	not required

7.4.2   **Identifying coordination entities and central atoms**

Within formulae, each coordination entity is enclosed within a set of square brackets. The first step is to identify each coordination entity for which a formula is to be created. The final formula for the compound is assembled by combining the formulae for the coordination entities with those of the other species present, using the standard rules of compositional nomenclature (Section 5.6). Within each coordination entity the central atom(s) must be identified.

7.4.3   **Identifying ligands**

Once a selection of the central atom(s) has been made, the remainder of the entity is considered to be made up of ligands.

7.4.4   **Selecting the representations of the ligands**

Each ligand must be represented either by an abbreviation or a sequence of element symbols that describes its composition. Formulae consisting of element symbols should be used for ligands that contain only a few atoms, so that, for example, water, ammonia, chloride, and nitrite can be represented by OH_2, NH_3, Cl, and NO_2, respectively. Where possible, the element symbol for the donor atom should be placed first (note that the order of the element symbols for each ligand can be changed to place the donor atom closer to the symbol for the central atom to which it is attached – see Section 7.4.6).

Abbreviations for ligand names are discussed in Section 7.4.1 above, and a selection of common abbreviations for use in formulae is presented in Table 7.1.

7.4.5   **Specifying donor atoms**

The donor atom is not normally specified unless there is something unusual about the coordination mode for a ligand. However, when a donor atom is not obvious, the donor atom(s) in a polyatomic ligand can be indicated within a formula by adding a κ-term to the end of the ligand representation and within the enclosing marks. The κ-term consists of the Greek letter κ followed immediately by the element symbol(s) for the donor atom(s). It is common practice to omit the κ-term for ligands that are well known to coordinate in one particular fashion. For example, -κO is not normally added to OH_2 as water usually coordinates to a single central atom through its oxygen atom. The ethane-1,2-▶ diamine ligand (abbreviated to 'en'), is most often bound in a bidentate fashion to a single central atom. The term -κN^1,N^2 added to 'en' to represent this coordination mode is normally not considered necessary. A bridging ligand can be indicated by adding the prefix 'μ' to the ligand representation.

7.4.6   **Assembling the formula**

The central atom symbol is placed first, followed by the ligands in alphabetical order, all within a pair of square brackets. The representation of a ligand determines the alphabetical order in which it is cited, so that NCMe or $NCCH_3$, which represent the same entity, may be placed in different positions within a formula, depending on which is selected. The third characters (M and C, respectively) are different and might result in different alphabetical ordering, depending on which other ligands are present. An N-bound cyanide should be distinguished from C-bound by changing the order of element symbols to place the donor atom symbol closer to the central atom to which it is bound (*e.g.*, NC compared with CN), and this will also affect the ordering of ligands within the formula.

The same representation should be used within a single document for all mentions of a ligand (unless the order is being altered to provide more structural information, for which see below). When determining the alphabetical order of ligands, the characters of an element symbol are treated as a single entity and each element is considered in order, so that, for example, CN (representing the cyanide ligand) would precede Cl, because C is compared to Cl.

Any ligand that contains more than one atom should be placed within enclosing marks. Any abbreviation that is chosen for a ligand must also be placed within enclosing marks. In the case of formulae, the nesting order for enclosing marks is different from that used in substitutive nomenclature (see Chapter 6), as square brackets are here reserved for enclosing coordination entities. The nesting order for formulae is therefore: [ ], [( )], [{( )}], [({( )})], [{({( )})}], *etc.* Multiple ligands of the same kind are indicated by adding the appropriate arabic numeral as a subscript, placed after the element symbol (for monatomic ligands) or after the second enclosing mark for polyatomic ligands.

If there is more than one central atom, the ordering rules should be altered to provide more structural information about the complex. Thus the formula for a dinuclear complex should place the representations for any bridging ligands between the symbols for the central atoms and the remaining ligand representations should be placed before the first central atom or after the second central atom, and closer to the central atom to which each ligand is attached. In these cases also, the element symbol sequence within a ligand representation should be changed to place the donor atom closer to the central atom to which it is attached (while still sensibly describing the structure of the ligand).

7.4.7   Describing specific isomers

It is often left to the reader to infer the geometry around the central atom in formulae. Generally one assumes that the donor atoms occupy positions as far apart as possible (*e.g.*, six ligands usually imply octahedral coordination, and four usually suggest tetrahedral coordination), but some central atoms almost invariably exhibit a preferred geometry, for example platinum(II) complexes are overwhelmingly square planar. The writer of a formula may not feel it necessary to specify the coordination geometry in such circumstances.

If there is an unusual geometry, or if a particular isomer of two or more needs to be identified, then an appropriate descriptor can be added to the front of the formula. Polyhedral symbols and configuration indices may be used for this purpose (see Sections 3.9 and 7.5.7), but in conjunction with formulae it is more common to use descriptors such as '*cis*' and '*trans*', where appropriate.

These steps in constructing a formula for a coordination entity are summarised in Figure 7.1, which should be used as a general guide to formula derivation, as exemplified in Table 7.2a.

7.5   NAMES OF COORDINATION COMPOUNDS

7.5.1   Introduction

The flow chart in Figure 7.2 provides a stepwise guide for the production of a name from a structural diagram (or other structural description) of a coordination entity. Examples 1–8 in Table 7.2b illustrate this process for mononuclear

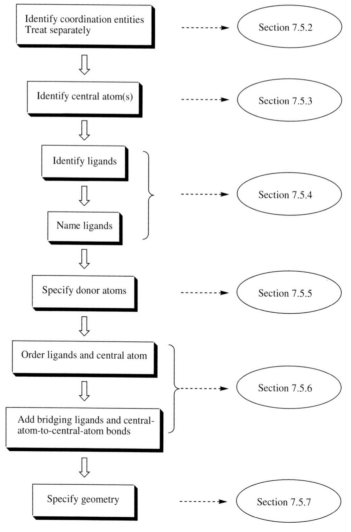

Figure 7.2 A flow chart showing the steps in producing a name from the structure of a coordination compound.

complexes, which are those with only one central atom, and for simpler dinuclear and polynuclear species with only one kind of central atom but all having the same coordination environment. The stages can be followed by moving down the chart in Figure 7.2. This is very like the flow chart Figure 7.1 described above, and the procedure outlined is very similar. The process for naming more complicated dinuclear and polynuclear complexes uses the same general approach, but the rules for describing the coordination mode for each ligand and the ordering of the components of the name are more complicated. These steps are described in the *Nomenclature of Inorganic Chemistry, IUPAC Recommendations 2005*.

7.5.2 Identifying Coordination Entities

The first step is the identification of the coordination entity to be named. In this context a coordination entity is any collection of atoms which can be treated as a central atom (or atoms) attached to one or more ligands by covalent bonds. If there is more than one such entity in a compound, then each should be treated separately (see Section 7.2.1, Example 2). The full name of the coordination compound is generally formed by combining the names of each coordination entity with those of any ions or other species present, using the standard rules of compositional nomenclature – essentially cations before anions, with other species listed after the anions, as cited in the discussion of salts, Section 5.6.

7.5.3 Identifying the Central Atom

Next the central atom of the complex must be identified. In most, but not in all, complexes the central atom is a metal ion. If there is more than one such atom, the complex is dinuclear or polynuclear. Non-metal atoms may also be central atoms, and this must be so when additive nomenclature is being used to describe compounds that do not contain a metal ion. Once again, if there is more than one central atom, the compound must be treated as dinuclear or polynuclear.

7.5.4 Identifying and Naming Ligands

Once the central atom(s) has been identified, the remainder of the coordination entity is treated as being made up of ligands. Each different kind of ligand must be identified and separately named according to the rules given below. Note that ligands can only be considered the same if they are bound to the central atom in the same way, through the same type of donor atom(s). The name for each ligand is used within the full additive name for the complex, except for names of anionic ligands, which are modified so that they end in 'o'. Thus the names of anionic ligands that end in 'ide', 'ate', or 'ite' are modified to end in 'ido', 'ato', or 'ito' when used in the names of complexes.

The bond between a central atom and a ligand is covalent, though the electron pair in such a bond is unlikely to be disposed symmetrically between both

Table 7.2b Examples of constructing names from structural formulae using additive (coordination) nomenclature. (As discussed in Section 7.1, in this text only, the symbol ▶ merely indicates a line break and is **NOT** a nomenclature symbol.)

Naming step	Example 1	Example 2	Example 3	Example 4
Structure to be named				
Identify central atom(s)	cobalt(III)	platinum(II)	cobalt(III)	sulfur
Identify and name ligands	ammonia → ammine water → aqua	chloride → chlorido triphenylphosphane	ethane-1,2-diyldinitrilotetra▶acetate → ethane-1,2-diyldinitrilotetraacetato	oxide → oxido peroxide → peroxido
Specify donor atom(s)	*not required for ammonia and water when using the ligand names above*	*not required for chloride* triphenylphosphane-κ*P*	ethane-1,2-diyldinitrilo▶-κ²*N*¹,*N*²-tetra(acetato-κ*O*)	*not required for oxide* peroxido-κ²*O*¹,*O*²
Assemble name	pentaammineaquacobalt(III) chloride	dichloridobis(triphenyl▶phosphane-κ*P*)platinum(II)	barium ethane-1,2-diyldinitrilo▶-κ²*N*¹,*N*²-tetra(acetato▶-κ*O*)cobalt(III)	μ-peroxido-κ²*O*¹,*O*²-bis▶(trioxidosulfate)(2−)
Describe coordina-tion geometry	*OC*-6	*SP*-4-1 or *trans*-	*OC*-6	*T*-4

Naming step	Example 5	Example 6	Example 7	Example 8
Structure to be named				
Identify central atom(s)	cobalt(III)	$2 \times$ rhenium	molybdenum	$4 \times$ iron
Identify and name ligands	2,3-dimethylbutane-2,3-diamine peroxide→peroxido	chloride→chlorido	chloride→chlorido 1,4,8,12-tetrathiacyclopentadecane	sulfide→sulfido methanethiolate→methanethiolato
Specify donor atom(s)	2,3-dimethylbutane-2,3-diamine-$\kappa^2 N^2,N^3$ η^2-peroxido	*not required for chloride*	*not required for chloride* 1,4,8,12-tetrathia-$\kappa^3 S^1,S^4,S^8$ ▲ -cyclopentadecane	*not required for sulfide* methanethiolato-κS
Assemble name	bis(2,3-dimethylbutane-2,3-diamine ▲ -$\kappa^2 N^2,N^3$)(η^2-peroxido)cobalt(III)	dicaesium bis(tetra ▲ chloridorhenate)(Re—Re)	trichlorido(1,4,8,12 ▲ -tetrathia-$\kappa^3 S^1,S^4,S^8$- ▲ cyclopentadecane)molybdenum(III)	tetra-μ_3-sulfido-tetrakis ▲ (methanethiolatoferrate)(2−)
Describe coordination and cluster geometry	*OC-6*	*SPY-5-12*	*OC-6-23*	*(T-4)-tetrahedro-*

partners. However, conventionally a ligand has been formally assumed to be in a closed shell configuration, and, for electron-counting purposes to donate an electron pair to the central atom. It is therefore a matter of convention whether a particular ligand is considered to be anionic or neutral. Ligated entities such as a chlorine atom or a methyl group are formally regarded as anionic when calculating central atom oxidation states. Ligated delocalised systems such as benzene, as well as a few odd-electron compounds such as NO, do not obey this two-electron formalism. Other potential ligands, such as phosphanes and amines, are always treated satisfactorily as two-electron donors.

A small number of common molecules are given special names when present in complexes. For example, a water ligand is represented in the full name by the term 'aqua', while 'ammine' represents an ammonia ligand (see Examples 1 in each of Sections 7.1 and 7.2, and Tables 7.2a and 7.2b). Carbon monoxide bound to the central atom through the carbon atom is represented by the term 'carbonyl' and nitrogen monoxide bound through nitrogen is represented by 'nitrosyl'. By convention, a single ligated hydrogen atom is always considered anionic and it is represented in the name by the term 'hydrido', whereas ligated dihydrogen is usually treated as a neutral two-electron donor species, and named 'dihydrogen'.

Ligands that contain carbon atoms (apart from cyanide and cyanate, thiocyanate and similar conventionally inorganic species) are considered organic molecules/ions, and are generally named using substitutive nomenclature (see Chapter 6, and Section 7.2, Examples 3 and 6, and Table 7.2b, Example 2). If the ligand is anionic, the name is then modified with an ending 'o' (see Section 7.2 Examples 1, 2, 3 and 5 and Table 7.2b, Example 3). Note that the name derived for a ligand using substitutive nomenclature will sometimes need to be modified, depending on the mode of coordination (compare Section 7.5.5, Example 2, and Table 7.2b, Example 3, and the two modes of binding shown in Section 7.5.5, Example 1.

Ligands that do not contain carbon are named using compositional nomenclature, *e.g.*, chloride, bromide, dioxygen, or acceptable short names that are widely used and recognised, *e.g.,* sulfate, nitrite, and thiocyanate. In more complex cases they themselves may be named as coordination entities, *e.g.*, $[O_3POPO_3]^{4-} = \mu$-oxido-bis(trioxidophosphate)(4–). Anionic ligand names are the anion names modified with the ending 'o', *e.g.*, 'chlorido', 'bromido', 'sulfato', 'nitrito', 'thiocyanato', and 'μ-oxido-bis(trioxidophosphato)(4–)'.

Bridging ligands bind to more than one central atom, and these are differentiated in names by the addition of the descriptor μ. This is separated from the name of the bridging ligand and from the rest of the name by hyphens (Section 7.2, Examples 4–6). This is sufficient if the bridging ligand is monatomic, but if the ligand is more complicated it may be necessary to specify which donor atom of the ligand is attached to which central atom. This is the case if the donor atoms are of different kinds, and κ-terms can then be used for this purpose (see Section 7.5.5).

7.5.5 **Specifying donor atoms**

A polyatomic ligand may be able to bind to a central atom in more than one way. It may use different donor atoms in different complexes. More than one atom may be bound to the central atom. More complicated situations are possible when there is more than one kind of central atom, but these are beyond the scope of this text. For details see the *Nomenclature of Inorganic Chemistry, IUPAC Recommendations 2005*. The name of a complex must allow for these different kinds of situation to be specified or described.

Specification of the donor atom can be achieved by adding a kappa-term (κ-term) to the name of the ligand. The κ-term includes the Greek letter κ and the italicised element symbol of the donor atom, and it may also be used to indicate to which central atom a ligand is bound if there is more than one kind. The *Nomenclature of Inorganic Chemistry, IUPAC Recommendations 2005*, discusses these possibilities in more detail. If the donor atoms are contiguous, then an eta-term (η-term) is used instead (see Chapter 8).

A κ-term is required for ligands where more than one coordination mode is possible. Typical cases are thiocyanate, which can be bound through either the sulfur atom (thiocyanato-κ*S*) or the nitrogen atom (thiocyanato-κ*N*), and nitrite, which can be bound through either the nitrogen atom (nitrito-κ*N*), or an oxygen atom (nitrito-κ*O*). The names pentaammine(nitrito-κ*N*)cobalt(III), Example 1(a), and pentaammine(nitrito-κ*O*)cobalt(III), Example 1(b), are used for each of the isomeric nitrito-complexes, and the κ-term is associated with the nitrito portion of the name to which it refers.

Example

1. Specifying different coordination modes

(a) nitrito-κ*N*   (b) nitrito-κ*O*

κ-Terms may be inserted within a ligand name, and if there is more than one donor atom in a ligand, more than one κ-term may be required. The coordination complex name in Example 2 contains, within the ligand name in the square brackets, two different κ-terms. The first κ-term indicates that the two different nitrogen atoms are coordinated to the platinum ion, the donor atoms being designated by the symbols *N* and *N'*. The second κ-term indicates that two acetate arms are also coordinated. The remaining two acetate arms in the ligand are not coordinated, and therefore do not require any κ-terms. In some cases, it may be necessary to add locants to make it clear that different donor atoms are meant (see Table 7.2b, Example 7).

Example

2. A complex in which the name of a molecule is changed because it is coordinated

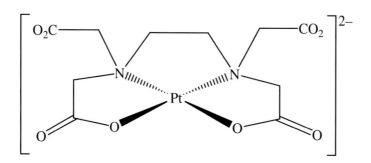

Name: [2,2′,2″,2‴-(ethane-1,2-diyldinitrilo-$\kappa^2 N$)tetraacetato-$\kappa^2 O^1,O^2$]platinate(2−)

Formula: [Pt(edta-$\kappa^4 N,N′,O,O′$)]$^{2-}$

If desired, a κ-term may be included in a formula, as in Example 2, but it is often impossible to insert a κ-term within a ligand abbreviation in any comprehensible way.

7.5.6   Assembling the name

Once the ligands have been named, and their coordination mode specified using any necessary κ-terms, the name can be assembled. If the complex has a single central atom, this is done by listing the ligand names in alphabetical order, and then adding the name of the central atom.

If there is more than one ligand of a particular kind bound to the central atom in the same way, the number of such identical ligands is indicated using a multiplicative prefix. For simple ligands, the prefixes 'di', 'tri', 'tetra', 'penta', 'hexa', *etc.* (Table P6), are used in the normal fashion. Any ligand that contains a substituent group, or has a name that contains enclosing marks (*e.g.*, parentheses), is not regarded as simple. Such ligands are placed within enclosing marks and the multiplicative prefixes (Table P6) 'bis', 'tris', 'tetrakis', 'pentakis', and 'hexakis', *etc.*, are used. The nesting order of enclosing marks, for use in names where more than one set of enclosing marks are required, is as follows: ( ), [( )], {[( )]}, ({[( )]}), *etc.*

If the overall charge on the complex is negative, the name of the central atom is modified by giving it the suffix 'ate'. Finally, the overall charge on the complex may be specified using an arabic numeral with the appropriate plus or minus sign, within parentheses. Alternatively, the oxidation state of the central atom can be added as Roman numerals, again within parentheses. It is generally not recommended to use both oxidation state and charge in the same name.

For the symmetrical dinuclear or polynuclear species that are considered here, the central atoms are all of one type (see Table 7.2b Examples 4, 6 and 8). These complexes have central atoms that are either bonded directly to each other (Example 6), or linked by bridging ligands (Examples 4 and 8). Such names can be constructed more simply than in cases where different types of central atom are involved. The process of naming complexes of this kind is similar to that for complexes with a single central atom. The difference is that a central atom and its non-bridging ligands are placed within enclosing marks and multiplicative prefixes are used to indicate how many repeating units are present. The names of bridging ligands are placed in front of the multiplicative prefix along with the accompanying μ-term.

Any repeating units linked by direct bonds between central atoms are indicated by placing italicised central atom symbols separated by a long ('em') dash in parentheses after the name of the repeating unit. The number (not bond order) of such bonds is indicated by an arabic numeral placed immediately before the first element symbol and separated from it by a space.

7.5.7 Describing coordination geometry

The approximate geometry around the central atom is described using a polyhedral symbol made up of letter codes for the geometry and a number that indicates the coordination number. Frequently used polyhedral symbols are shown in Table P5. The polyhedral symbol is placed before the name of the complex.

The relative positions of donor atoms around a central atom can be described using a configuration index (a sequence of numbers that follows the polyhedral symbol, and if required, the absolute configuration can also be described (Section 3.9). These are only necessary if there is more than one possibility and a particular stereoisomer needs to be identified.

Alternatively, a range of trivial isomer descriptors may be used in specific limited situations. For example, the two geometrical isomers possible when a square planar centre is coordinated by two donor atoms of one type and two of another are referred to as *cis* (when the identical donors are coordinated next to each other) or *trans* (when the identical donors are coordinated opposite to each other). Octahedral centres with four donor atoms of one kind and two of another can also be referred to as *cis* (when the distinctive ligands are coordinated next to each other) or *trans* (when they are coordinated opposite each other). Octahedral centres with three of each of two kinds of donor atom can be described as *mer* (or meridional, when the three donor atoms of each kind are located on the meridian of the octahedron, *i.e.*, when two are coordinated opposite to each other and the other lies between) or *fac* (or facial, when the three donor atoms of a particular kind are located at the corners of a triangular face of the octahedron). These descriptors can only be applied to the more simple cases, such as those cited here.

If three or more central atoms form a cluster, a structural descriptor should be added as a prefix to the name of the cluster to describe the spatial relationship between the central atoms. These descriptors can be found in Table 7.3, and the use is exemplified in Table 7.2b, Example 8.

Table 7.3 Structural descriptors for small clusters of central atoms

Number of central atoms	Spatial arrangement	Point group	Descriptor
3	Triangular	D_{3h}	*triangulo*
4	Square	D_{4h}	*quadro*
4	Tetrahedral	T_d	*tetrahedro*
6	Octahedral	O_h	*octahedro*
6	Trigonal prismatic	D_{3h}	*triprismo*

7.6 DECIPHERING NAMES AND FORMULAE

Names and formulae must be of a form that can be interpreted by a reader, who should then be able to produce a structural diagram. Clearly such a process is related to the steps presented in the flow charts (Figures 7.1 and 7.2) for the production of names and formulae. Deciphering a name or formula can be approached in a number of different ways. A general approach is described in Chapter 13.

REFERENCE

1. Graphical representation of stereochemical configuration, Recommendations 2006, *Pure and Applied Chemistry*, 2006, **78**(10), 1897–1970; Graphical representation standards for chemical structure diagrams, Recommendations 2007, *Pure and Applied Chemistry*, 2008, **80**(2), 277–410.

8 Nomenclature of Organometallic Compounds

8.1 INTRODUCTION

An organometallic compound is any compound containing at least one bond between a metal atom and a carbon atom. The nomenclature of organometallic compounds of the transition elements is based very heavily on that of coordination compounds, but the enormous growth in organometallic chemistry over the last fifty years has uncovered new classes of compound with unprecedented bonding modes and has resulted in the need for additional nomenclature rules for organometallic compounds.

The major part of this Chapter outlines a system for naming transition element organometallic compounds based on the additive nomenclature as introduced in Section 4.4 and applied to coordination compounds in Chapter 7, but incorporating, as far as possible, rules for naming organic ligands (reference 1). In addition, further rules are formulated to designate unambiguously the special modes of bonding often found in organometallic compounds.

The last part of this Chapter describes briefly aspects of the naming of Main Group organometallic compounds, where substitutive nomenclature (presented in Chapter 6) is applied by substituting the appropriate parent hydrides of the elements of Groups 13–16. The names of organometallic compounds of Group 1 and 2 elements are derived using additive nomenclature.

Just as in Chapter 7 on coordination compounds, the discussion in this Chapter is limited to the nomenclature of compounds with only one type of central atom in each coordination entity. For a full treatment of polynuclear complexes, especially where the central atoms are of different elements, the reader is referred to the *Nomenclature of Inorganic Chemistry, IUPAC Recommendations 2005* (reference 2). That book also provides a wealth of further examples to illustrate the nomenclature described in the present text.

8.2 ORGANOMETALLIC COMPOUNDS OF THE TRANSITION ELEMENTS

8.2.1 General

Organometallic compounds of the transition metals are named as coordination compounds, using the additive nomenclature system, and following the steps shown in the flow chart of Figure 7.2. It is assumed that the reader of the current Chapter has an understanding of the principles of the additive nomenclature of

Principles of Chemical Nomenclature: A Guide to IUPAC Recommendations
Edited by G. J. Leigh
© International Union of Pure and Applied Chemistry 2011
Published by the Royal Society of Chemistry, www.rsc.org

coordination compounds presented in Chapter 7. If the ligand is organic but it coordinates to a central atom *via* an atom other than carbon, the name of the ligand is developed using substitutive nomenclature, and the usual rules for naming ligands in coordination entities are applied. However, additional rules are required for the step in the flow chart (Figure 7.2) where ligands bonding to the metal through carbon are named, and the special nature of the bonding of unsaturated hydrocarbons to metals via their π-electrons requires the introduction of the eta (η) convention to specify the number of contiguous ligating atoms (Section 8.2.5.1).

8.2.2 **Compounds with one metal-carbon single bond**

Organic ligands coordinating *via* one carbon atom may be regarded as anions formed by the removal of one hydron (see Section 2.2) from a carbon atom of an organic molecule. These anions are named by replacing the final 'e' of the parent compound name by 'ide'. Thus methane becomes methanide. As in the nomenclature of coordination compounds, the suffix 'ide' is then replaced by 'ido' upon coordination of the ligand. Thus methanide becomes methanido upon coordination. The 'anionic' atom is given a locant as low as is consistent with any established numbering of the parent hydride, and the locant number, including '1', must always be cited for linear hydrocarbons with more than two carbon atoms (though not for monocyclic, unsubstituted rings). However, note that fused polycyclic hydrocarbons and heterocyclic systems have special numbering schemes (see Section 6.3). Applying these rules, the compound [TiCl$_3$Me] is called trichlorido(methanido)titanium. Note that the parentheses help to distinguish this from a name such as trichloromethanidotitanium, [Ti(CCl$_3$)], though this last species is hypothetical and certainly unstable.

However, it has long been the custom to call the ligand CH$_3$ methyl, often abbreviated to Me. This name is used for CH$_3$ in substitutive nomenclature. In both the older and the recent literature the name methyl has also been used in additive nomenclature. Of course, though the designation of such ligands in organometallic chemistry as anionic rather than neutral is somewhat arbitrary, the metal-carbon bond may be highly covalent. In spite of the established use of the ligand name 'methyl' in organic and organometallic chemistry, [TiCl$_3$Me] may now receive the name 'trichloridomethyltitanium' as well as 'trichloridomethanidotitanium'. Similar reasoning could be applied to all related ligands when additive nomenclature is employed. The current prevalence of the older, alternative substitutive names means that these must be considered when interpreting names found in both the older and current literature, so reference 2, for example, allows the use of both 'methyl' and 'methanido' for the ligand CH$_3$. Similarly, both classes of name are currently allowed for all related organic ligands. Consistency of naming style should be observed in any piece of written work.

Table 8.1 gives the names used for typical ligands forming a single bond to a metal, and this is followed by examples illustrating the names of compounds containing one metal-carbon single bond. In this Table and in Tables 8.2 and 8.3 the organic anion ligand names are listed as well as the traditional names applied to the same groups as substituents and also used as ligand names. These

Table 8.1 Names for ligands forming a metal-carbon single bond

Ligand formula	Systematic anionic ligand name	Frequently encountered substitutive name	Alternative name still sometimes employed
CH_3-	methanido	methyl	
CH_3CH_2-	ethanido	ethyl	
$CH_3CH_2CH_2-$	propan-1-ido	propyl	
$(CH_3)_2CH-$	propan-2-ido	propan-2-yl **or** 1-methylethyl	isopropyl
$CH_2=CHCH_2-$	prop-2-en-1-ido	prop-2-en-1-yl	allyl
$CH_3CH_2CH_2CH_2-$	butan-1-ido	butyl	
$CH_3CH_2-\overset{CH_3}{\underset{H}{C}}-$	butan-2-ido	butan-2-yl **or** 1-methylpropyl	*sec*-butyl
$\underset{H_3C}{\overset{H_3C}{>}}CH-CH_2-$	2-methylpropan-1-ido	2-methylpropyl	isobutyl
$H_3C-\overset{CH_3}{\underset{CH_3}{C}}-$	2-methylpropan-2-ido	2-methylpropan-2-yl **or** 1,1-dimethylethyl	*tert*-butyl
cyclobutane-CH— structure	cyclobutanido	cyclobutyl	
C_5H_5-	cyclopenta-2,4-dien-1-ido*	cyclopenta-2,4-dien-1-yl	cyclopentadienyl
C_6H_5-	benzenido	phenyl	
$C_6H_5CH_2-$	phenylmethanido	phenylmethyl	benzyl
$H_3C-\overset{O}{C}\backslash$	1-oxoethan-1-ido	ethanoyl	acetyl
phenyl-C(=O)— structure	oxo(phenyl)methanido	benzenecarbonyl	benzoyl
naphthalen-2-yl structure	naphthalen-2-ido	naphthalen-2-yl	2-naphthyl

*Although strictly speaking ambiguous, the ligand name cyclopentadienido is acceptable as a short form of cyclopenta-2,4-dien-1-ido (Section 8.2.5.1).

substituent names are widely found in the literature. The alternative names given in the final column are also frequently encountered. The names in Examples 1–6 below are constructed according to the principles of the nomenclature of coordination compounds outlined in Chapter 7 and using the flow chart in Figure 7.2. The first name given is that obtained when considering the hydrocarbon ligand as an anionic species; in addition, the widely encountered variant is also illustrated.

Examples

1. [OsEt(NH$_3$)$_5$]Cl
   pentaammine(ethanido)osmium(1+) chloride **or**
   pentaammine(ethyl)osmium(1+) chloride

2. Li[CuMe$_2$]
   lithium dimethanidocuprate(1−) **or**
   lithium dimethylcuprate(1−)

3. [Pt{C(O)Me}Me(PEt$_3$)$_2$]
   methanido(1-oxoethan-1-ido)bis(triethylphosphane)platinum **or**
   acetyl(methyl)bis(triethylphosphane)platinum

4.

   (phenylethynido)(pyridine)bis(triphenylphosphane)rhodium **or**
   (phenylethynyl)(pyridine)bis(triphenylphosphane)rhodium

5.

   P⌒P = Me$_2$PCH$_2$CH$_2$PMe$_2$

      = ethane-1,2-diyl-bis(dimethylphosphane)

   bis[ethane-1,2-diylbis(dimethylphosphane-κ*P*)]hydrido(naphthalen-2-▶
   ido)ruthenium **or**
   bis[ethane-1,2-diylbis(dimethylphosphane-κ*P*)]hydrido(naphthalen-2-▶
   yl)ruthenium

6.

   carbonyl(η^5-cyclopentadienido)[(*E*)-3-phenylbut-2-en-2-▶
   ido](triphenylphosphane)iron **or**
   carbonyl(η^5-cyclopentadienyl)[(*E*)-3-phenylbut-2-en-2-▶
   yl](triphenylphosphane)iron

Note that in Example 6 the names contain the symbol η, the significance of which is discussed in Section 8.2.5.1, and also the designator (*E*), meaning *trans*, as discussed in Section 3.9.2.

8.2.3 Compounds with several metal-carbon single bonds from one ligand

8.2.3.1 *General.* When an organic ligand forms two or three metal-carbon single bonds (to one or more metal atoms), the ligand may be treated as a di- or tri-anion, with the suffixes 'diido' or 'triido' being used, with no removal of the terminal 'e' of the name of the parent hydrocarbon. In the conventional fashion, the position '1' is assigned so as to define the longest chain of carbon atoms, and the direction of numbering is chosen to give the lowest possible locants to the side-chains or substituent positions. The locant(s) must always be cited. This nomenclature also applies to hypervalent coordination modes, *e.g.* for bridging methanido groups. Other names for such ligands, those derived by regarding them as if they were substituent groups and using the suffixes 'diyl' and 'triyl' attached to the name of the parent hydrocarbon, are still commonly encountered.

Typical ligands forming two or three metal-carbon single bonds are listed in Table 8.2, where alternative forms to be found in the literature are also listed.

8.2.3.2 *Bridging ligands and metal–metal bonds.* Organic ligands forming more than one metal-carbon bond may be either chelating, if coordinating to a single metal atom (Example 7), or bridging, if coordinating to two or more metal atoms (Example 8). A bridging bonding mode is indicated by a **bridging index**, the Greek letter μ (mu) prefixing the ligand name (see Section 7.2.5 for an earlier discussion of this index).

Table 8.2 Names for ligands forming several metal-carbon single bonds

Ligand formula	Systematic anionic ligand name	Frequently encountered substitutive name	Alternative name still sometimes employed
$-CH_2-$	methanediido	methanediyl	methylene
$-CH_2CH_2-$	ethane-1,2-diido	ethane-1,2-diyl	ethylene
$-CH_2CH_2CH_2-$	propane-1,3-diido	propane-1,3-diyl	
$HC-$	methanetriido	methanetriyl	
CH_3HC	ethane-1,1-diido	ethane-1,1-diyl	
CH_3C-	ethane-1,1,1-triido	ethane-1,1,1-triyl	
$-CH=CH-$	ethene-1,2-diido	ethene-1,2-diyl	
$H_2C=C$	ethene-1,1-diido	ethene-1,1-diyl	
$-C\equiv C-$	ethyne-1,2-diido	ethyne-1,2-diyl	acetylido
$-C_6H_4-$	benzene-1,2-diido (**or** -1,3- **or** -1,4-, depending on the isomer)	benzene-1,2-diyl (**or** -1,3- **or** -1,4-, depending on the isomer)	1,2-phenylene (**or** 1,3- **or** 1,4-, depending on the isomer)

Examples

A chelating ligand

A bridging ligand

7. 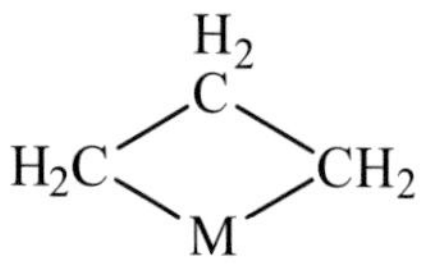 $H_2C\overset{\overset{\displaystyle H_2}{C}}{\diagdown}CH_2$ — M (chelating)

8. $H_2C\overset{\overset{\displaystyle H_2}{C}}{\diagup\diagdown}CH_2$ — M   M (bridging)

propane-1,3-diido **or**
propane-1,3-diyl

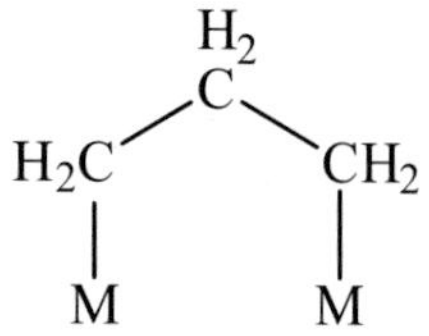

μ-propane-1,3-diido **or**
μ-propane-1,3-diyl

The number of metal atoms connected by a bridging ligand is indicated by a right subscript, μ_n, where $n \geq 2$, though when the bridging index is 2 it is not normally stated and the symbol μ is then used without qualification (compare Examples 9 and 10).

Examples

9. 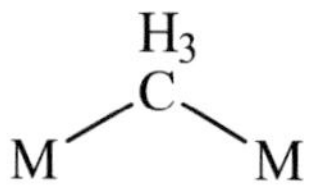 $M\overset{\overset{\displaystyle H_3}{C}}{\diagup\diagdown}M$

10. $M\overset{\overset{\displaystyle H_3}{C}}{\diagdown\mid\diagup}M$ (with M below)

μ-methanido **or**
μ-methyl

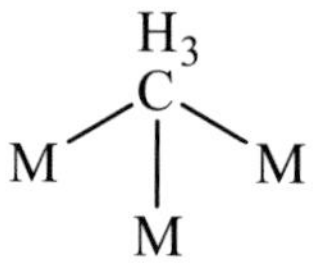

μ₃-methanido **or**
μ₃-methyl

Note that the name methylene for CH_2 can only be used in connection with a bridging bonding mode (μ-methylene, Example 11), whereas a CH_2 ligand bonding to a single metal has a metal-carbon double bond and should be named as methylidene (Example 12) (see Section 8.2.4 for a discussion of compounds with multiple carbon-metal bonds).

Examples

11. 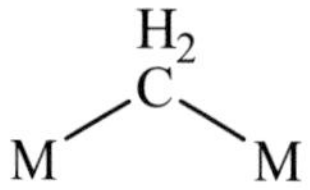 $M\overset{\overset{\displaystyle H_2}{C}}{\diagup\diagdown}M$

12. $M{=}CH_2$

μ-methylene

methylidene

The group CH_2CH_2 is μ-ethane-1,2-diido when bridging (Example 13) but CH_2CH_2 coordinating through both carbon atoms to a single metal centre is η^2-ethene (Example 14) (see Section 8.2.5.1 for a discussion of the η symbol and its use).

Examples

13. 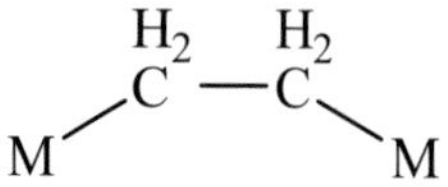 $M\overset{\overset{\displaystyle H_2}{C}}{\diagup}{-}\overset{\overset{\displaystyle H_2}{C}}{\diagdown}M$

14. $H_2C{=}CH_2$ — M

μ-ethane-1,2-diido **or**
μ-ethane-1,2-diyl

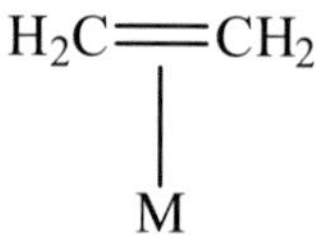

η^2-ethene

A similar situation arises with CHCH which, when bridging with the carbon atoms individually singly bonded to each of two metals, is called μ-ethene-1,2-▶ diido (Example 15) but when the metal-carbon bonds are double, μ-ethanediylidene (Example 16). When this same grouping is coordinated through both carbon atoms to one metal atom it is η^2-ethyne (Example 17), but if both carbon atoms are coordinated to both metal centres it is necessary to use both symbols, η and μ to describe it, and the name becomes μ-η^2:η^2-ethyne (Example 18). Students should be aware that it is sometimes not obvious whether a given structure is represented best by a diagram such as in Example 15, or such as in Example 16. In such circumstances, individual judgement may need to be exercised.

Examples

15.

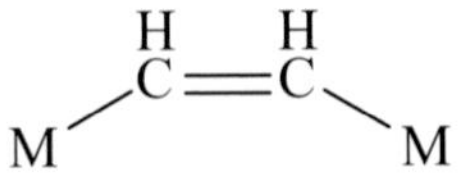

$$\begin{array}{ccc} & H \quad\quad H & \\ & C = C & \\ M^{\diagup} & & {}^{\diagdown}M \end{array}$$

μ-ethene-1,2-diido, **or**
μ-ethene-1,2-diyl

16.

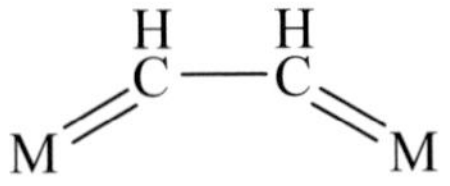

$$\begin{array}{ccc} & H \quad\quad H & \\ & C - C & \\ M^{\diagup\diagup} & & {}^{\diagdown\diagdown}M \end{array}$$

μ-ethanediylidene

17. HC≡CH
|
M

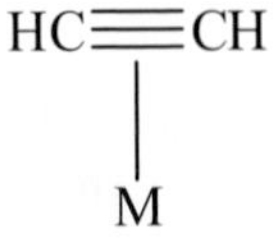

η^2-ethyne

18. HC≡CH

M—M

μ-η^2:η^2-ethyne

The usual formalisms apply to name construction. Bridging ligands are listed in alphabetical order as are the other ligands, but in names a bridging ligand is cited before a corresponding non-bridging ligand. Multiple bridging is listed in decreasing order of complexity, *e.g.* μ_3 bridging being cited before μ_2 bridging.

Bridging ligands often span central atoms between which metal–metal bonding occurs (see Section 7.2.6). Metal–metal bonding is indicated by the italicised element symbols of the appropriate metal atoms, separated by an 'em' dash and enclosed in parentheses, placed after the list of central atoms and before the ionic charge. The number of such metal–metal bonds is indicated by an arabic numeral placed before the first element symbol and separated from it by a space. For the purpose of nomenclature, no distinction is made between different metal–metal bond orders.

Examples

19.

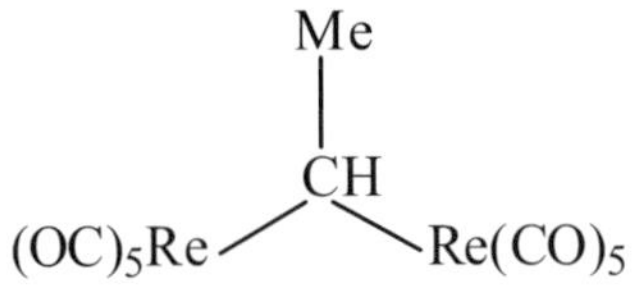

$$\begin{array}{c} Me \\ | \\ CH \\ (OC)_5Re^{\diagup} \quad {}^{\diagdown}Re(CO)_5 \end{array}$$

(μ-ethane-1,1-diido)bis(pentacarbonylrhenium) **or**
(μ-ethane-1,1-diyl)bis(pentacarbonylrhenium)

20.

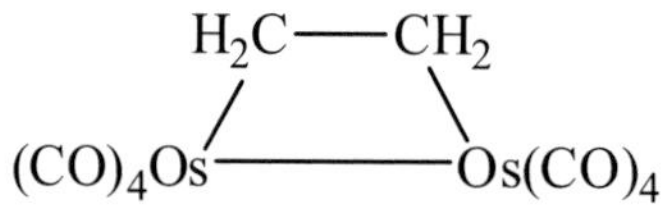

$$H_2C \text{——} CH_2$$

(CO)$_4$Os ———— Os(CO)$_4$

(μ-ethane-1,2-diido)bis(tetracarbonylosmium)(*Os—Os*) **or**
(μ-ethane-1,2-diyl)bis(tetracarbonylosmium)(*Os—Os*)

21.

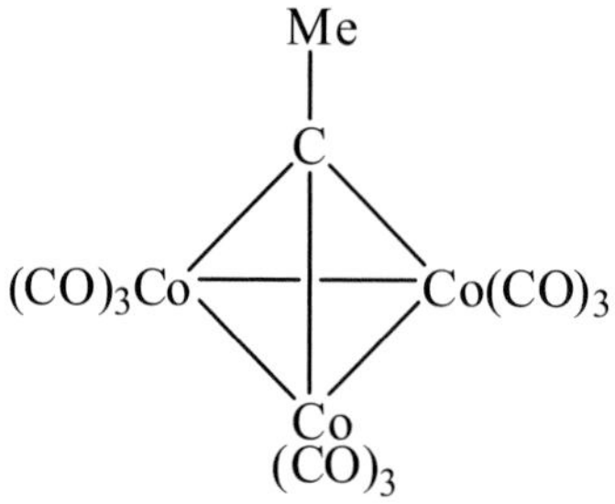

Me
|
C

(CO)$_3$Co ——|—— Co(CO)$_3$

Co
(CO)$_3$

(μ$_3$-ethane-1,1,1-triido)-*triangulo*-tris(tricarbonylcobalt)(3 *Co—Co*) **or**
(μ$_3$-ethane-1,1,1-triyl)-*triangulo*-tris(tricarbonylcobalt)(3 *Co—Co*)

For a more detailed discussion of ligand bridging modes in dinuclear compounds and in larger polynuclear clusters, including those with different kinds of central atom, see Section IR-9.2.5 in reference 2.

8.2.3.3   *Chelating ligands.* Where a chelating ligand is formed by removing two or more hydrons from a parent compound, the 'anionic' atoms, understood to form the bonds to the central atoms, are indicated by using the appropriate locants in the ligand name (Examples 22–24).

Examples

22.

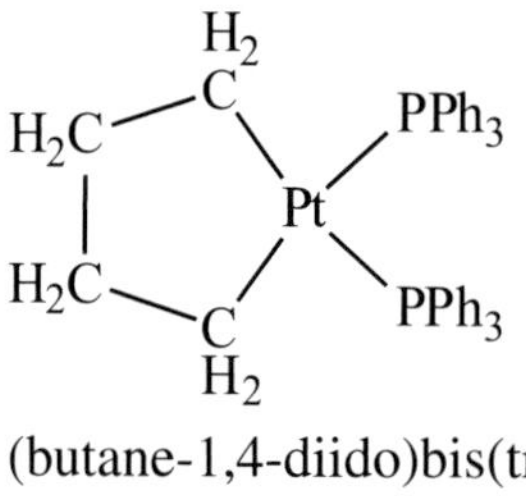

H$_2$
C
H$_2$C   PPh$_3$

H$_2$C   Pt

H$_2$C   PPh$_3$
C
H$_2$

(butane-1,4-diido)bis(triphenylphosphane)platinum **or**
(butane-1,4-diyl)bis(triphenylphosphane)platinum

23.

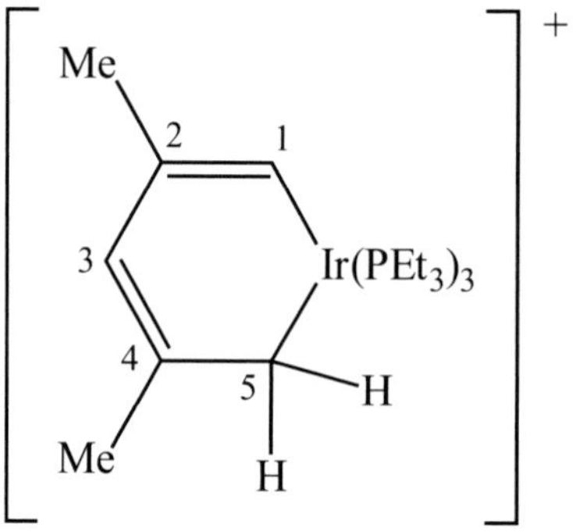

[Me

2   1

3   Ir(PEt$_3$)$_3$

4
5   H
Me   H]$^+$

(2,4-dimethylpenta-1,3-diene-1,5-diido)tris(triethylphosphane)iridium(1+) **or**
(2,4-dimethylpenta-1,3-diene-1,5-diyl)tris(triethylphosphane)iridium(1+)

24.

(1-oxo-2,3-diphenylpropane-1,3-diido)bis(triphenylphosphane)platinum **or**
(1-oxo-2,3-diphenylpropane-1,3-diyl)bis(triphenylphosphane)platinum

When a chelate ring contains a coordinate (formerly also called a dative, or semi-polar) bond from a heteroatom in addition to a carbon attachment, the ligand should be named using the κ convention. In this convention, which is introduced in Sections 7.4.5 and 7.5.5, the coordinating atoms of a polydentate ligand bonding to a metal centre are indicated by the Greek letter kappa, κ, preceding the italicised element symbols of the ligating atoms. A right superscript numeral may be added to the symbol κ to indicate the number of identically bound ligating atoms; non-equivalent ligating atoms which differ from each other, either in kind or in bonding, should each be indicated by an italicised element symbol. The designator η (Section 8.2.5.1) is treated similarly when two or more different sets of contiguous ligating atoms are considered (Example 49).

In simple cases one or more superscript primes on the italicised element symbol may be used to differentiate donor atoms of the same element. Otherwise a numerical right superscript corresponding to the conventional numbering of the atoms in the ligand is used to define unambiguously the identity of the ligating atom. As shown in the Examples below, these symbols are placed *after* that portion of the ligand name which represents the particular functionality, substituent group, ring, or chain in which the ligating atom is found.

Often it is only necessary for the coordinating heteroatom to be specified using the κ convention, the ligating carbon atom being adequately specified by the appropriate substitutive suffix. The use of an arrow to indicate a coordinate (semi-polar or dative) bond as shown in Examples 25 and 26 was common but is no longer encouraged. In Example 25 no term κC^1 to designate the bonding of the phenyl group to the metal is included because the name itself implies that this must be from carbon atom number 1.

Examples

25.

tetracarbonyl[2-(2-phenyldiazen-1-yl-κN^2)benzen-1-ido-κC^1]manganese **or**
tetracarbonyl[2-(2-phenyldiazen-1-yl-κN^2)phenyl]manganese

26.

chloridohydrido(2-methyl-3-oxo-κO-but-1-en-1-ido-κC¹)bis(triisopropylphosphane)rhodium **or**
chloridohydrido(2-methyl-3-oxo-κO-but-1-en-1-yl)bis(triisopropylphosphane)rhodium

8.2.4   Compounds with metal-carbon multiple bonds

Many organometallic compounds contain a multiple metal-carbon bond, such as the well-known first-generation Grubbs' catalyst for olefin metathesis, Example 27, which has a double bond between ruthenium and carbon. Whereas the anion names given in Table 8.2 (methanediido, methanetriido, *etc.*) could be used for ligands forming metal-carbon double or triple bonds, this nomenclature does not necessarily imply the concept of the metal-ligand bond being double or triple. The terms 'carbene' and 'carbyne', whilst in common use to describe entities able to form metal-carbon double or triple bonds, are not suitable for nomenclature purposes. For example, 'carbene' properly refers to the free R_2C: species. When these are ligands they are given substituent prefix names derived from the parent hydrides (*cf.* Section 6.4.2) which end with the suffix 'ylidene' for a double metal-carbon bond and with the suffix 'ylidyne' for a triple bond. These suffixes either replace the ending 'ane' of the parent hydride name with 'ylidene' or 'ylidyne' or, more generally, are added to the name of the parent hydride with elision of the terminal 'e', if present. Thus the entity $CH_3CH_2CH=$ as a ligand is named propylidene and $(CH_3)_2C=$ is called propan-2-ylidene.

Note that if a ligand forms one or more metal-carbon single bonds as well as metal-carbon multiple bonds, the order of suffixes is 'yl', or 'ylidene' 'ylidyne'. Further details on the use of these suffixes and a list of names of typical ligands forming metal-carbon double or triple bonds may be found in Section IR-10.2.4 of reference 2.

Examples

27.

dichlorido(phenylmethylidene)bis(tricyclohexylphosphane)ruthenium **or**
(benzylidene)dichloridobis(tricyclohexylphosphane)ruthenium

28.

(2,4-dimethylpenta-2,4-diene-1,1,5-triido-$\kappa^2 C^1,C^5$)tris(triethylphosphane)iridium
or (2,4-dimethylpenta-1,3-dien-1-yl-5-ylidene)tris(triethylphosphane)iridium

29.

tetracarbonyl[(diethylamino)methylidyne]iodidochromium

8.2.5   Compounds containing bonds to unsaturated molecules or groups

8.2.5.1   *The eta (η) convention.* Since the discovery of Zeise's salt, K[Pt(η^2-C_2H_4)Cl$_3$], the first organometallic complex of a transition element, and particularly since the first reported synthesis of ferrocene, [Fe(η^5-C_5H_5)$_2$], the number and variety of organometallic compounds containing bonds to unsaturated organic ligands has increased enormously and examples now pervade the literature. The use of the symbol η to describe such bonding is discussed here.

Complexes containing ligands which coordinate to a central atom with at least two adjacent atoms in a 'side-on' fashion require a special nomenclature. Entities such as alkenes, alkynes and aromatic compounds contain groups that may coordinate *via* the π-electrons of their multiple bonds, but there are also carbon-free entities containing bonds involving heteroelements which can similarly coordinate. Such compounds are generally referred to as π-complexes, but the use of π and σ in nomenclature is not recommended. The symbols π and σ really refer to the symmetries of orbitals and their interactions and are effectively quantum numbers, which are irrelevant for nomenclature purposes.

The special nature of the bonding of unsaturated hydrocarbons to metals *via* their π-electrons led to the development of the **'hapto' nomenclature**, introduced in its original form by F. A. Cotton in 1968, to designate unambiguously the unique bonding modes of the compounds so formed (reference 3). In this **'eta' convention**, the Greek symbol η (eta) is used to indicate the nature of the connectivity between the ligand and the central atom. The number of *contiguous* atoms in the ligand coordinated to the metal (the **hapticity** of the ligand) is indicated by a right superscript, *e.g.* η^3 ('eta three' or 'trihapto'), η^4 ('eta four' or 'tetrahapto'), η^5 ('eta five' or 'pentahapto'), *etc.* The symbol η is added as a prefix to the ligand name, or to that portion of the ligand name most appropriate, to indicate the connectivity, as in 5-(η^2-ethenyl)cyclopenta-1,3-diene ethene (Example 30) and ethenyl-η^5-cyclopentadienido (Example 31).

Examples

30. 5-(η^2-ethenyl)cyclopenta-1,3-diene

31. ethenyl-η^5-cyclopentadienido

The ligand name η^5-cyclopentadienido is ambiguous, strictly speaking, but it is acceptable as a short form of η^5-cyclopenta-2,4-dien-1-ido. In addition, the alternative name, η^5-cyclopentadienyl, will be found extensively in the literature.

These ligand names are enclosed in parentheses in the full name of a complex. Note the importance of making rigorous use of enclosing marks to distinguish the above bonding modes (Examples 30 and 31) from the other four cases shown in Examples 32–35. Note also that when cyclopenta-2,4-dien-1-ido coordinates by the carbon with the free valence (*i.e.* via a σ bond), a κ term is added for explicit indication of that bonding. The symbol η^1 should never be used in this context as **the eta convention applies only to the bonding of contiguous atoms** in the ligand. Attachment of a ligand through a single atom, or through several non-contiguous atoms, must be specified using the κ nomenclature. This differentiation may be necessary in the names of unsaturated ligands which may participate in several types of bonding.

Examples

32. (cyclopenta-2,4-dien-1-ido-κC^1)(η^2-ethene)

33. (η^5-cyclopentadienido)(η^2-ethene)

34. (cyclopenta-2,4-dien-1-ido-κC^1)(ethenido)

35. (η^5-cyclopentadienido)(ethenido)

Note that to designate a π-bonded system diagrams of both types depicted below are currently used.

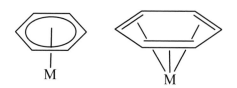

The structures of metal complexes of unsaturated ligands and the nature of the bonding in them is sometimes complicated or ill-defined. Further, such ligands may be regarded formally as anionic or neutral (or sometimes even cationic). Consequently, names for π-bonding ligands are chosen so that they indicate stoichiometric composition and are derived in a similar way to those for the ligands discussed in preceding Sections. Therefore, unsaturated ligands considered to be neutral molecules, such as alkenes, alkynes, nitriles and diazenes, including ones with more than one unsaturated unit, such as buta-1,3-diene (C_4H_6), benzene (C_6H_6), norborna-2,5-diene or bicyclo[2.2.1]hepta-2,5-diene (C_7H_8), cyclohepta-1,3,5-triene (C_7H_8), cycloocta-1,5-diene (C_8H_{12}) and cycloocta-1,3,5,7-tetraene (C_8H_8), are given names in accordance with the rules of Chapter 6. This name is prefixed by the appropriate eta symbol, η^n, to indicate the hapticity (n) of the ligand, which for a particular ligand may vary from compound to compound. For example, cyclooctatetraene has the potential to display coordination modes of η^2, η^4, η^6 and η^8, and in binuclear complexes η^3 and η^5 are also found.

When these π-bonding ligands are in the form of anions obtained by removing hydrons from (substituted) parent hydrides, their names are given the suffixes 'ido', 'diido', *etc.*, depending on the number of hydrons removed. This name (or the most appropriate portion of the name to indicate the connectivity) is then prefixed by the appropriate eta symbol. Names obtained by treating the ligands as substituent groups derived by removing hydrogen atoms from (substituted) parent hydrides and given the substituent names ending in 'yl', 'diyl', *etc.*, will also be found in the literature. Table 8.3 lists names for a selection of these unsaturated species acting as ligands. Since there is little likelihood of confusion, shortened forms of anion and substituent group names are often acceptable, *e.g.*, η^5-cyclohexadienido instead of η^5-cyclohexa-2,4-dien-1-ido and η^5-cyclohexadienyl instead of η^5-cyclohexa-2,4-dien-1-yl.

Complexes of unsaturated systems incorporating heteroatoms may be designated in the same manner if both the carbon atoms and adjacent heteroatoms are coordinated, and the eta convention may also be extended to π-coordinated ligands containing no carbon atoms, such as cyclotriborazane and pentaphosphole.

The flow chart in Figure 7.2 using the eta convention was employed for naming the π-bonded ligands in the following Examples 36–42. Where two names are given, the first is that obtained when considering the unsaturated ligand as an anionic species; where appropriate, the commonly found alternative name derived as if the ligand were a neutral substituent group is also illustrated.

Examples

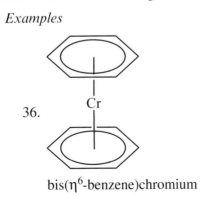

36.

bis(η^6-benzene)chromium

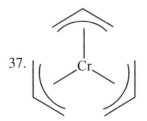

37.

tris(η^3-propenido)chromium **or**
tris(η^3-propenyl)chromium **or**
tris(η^3-allyl)chromium

Table 8.3 Names for selected unsaturated ligands

Ligand formula*	Systematic anionic ligand name	Frequently encountered substitutive name	Alternative name still sometimes employed
	η^3-propenido	η^3-propenyl	η^3-allyl
	η^5-pentadienido	η^5-pentadienyl	
	η^5-cyclopentadienido	η^5-cyclopentadienyl	
	pentamethyl-η^5-cyclopentadienido	pentamethyl-η^5-cyclopentadienyl	
	η^5-cyclohexadienido	η^5-cyclohexadienyl	
	η^7-cyclooctatrienido	η^7-cyclooctatrienyl	
	η^5-azacyclopentadienido	η^5-azacyclopentadienyl	η^5-pyrrolyl

*The ligands are drawn as if complexed to a metal, *i.e.*, these are depictions of bonded entities, not free ligands. The arcs used in these and other examples indicate electron delocalisation, analogous to the formalism used for benzene.

38.

dicarbonyl(η^5-cyclopentadienido)(cyclopenta-2,4-dien-1-ido-κC^1)iron **or**
dicarbonyl(η^5-cyclopentadienyl)(cyclopenta-2,4-dien-1-yl-κC^1)iron

39.

dicarbonyl(η^3-cyclopentadienido)(η^5-cyclopentadienido)tungsten **or**
dicarbonyl(η^3-cyclopentadienyl)(η^5-cyclopentadienyl)tungsten

40.

tricarbonyl{1-[2-(diphenylphosphanyl)-η^6-phenyl]-*N*,*N*-dimethyl▶
ethan-1-amine}chromium

41.

(η^2-carbon dioxide)bis(triethylphosphane)nickel

42.

(pentamethyl-η^5-cyclopentadienido)(η^5-pentaphospholido)iron **or**
(pentamethyl-η^5-cyclopentadienyl)(η^5-pentaphospholyl)iron

123

If not all unsaturated atoms of a ligand are involved in bonding, or if a ligand can adopt several bonding modes, or if a ligand bridges several metal atoms, the locants of the ligating atoms must be cited in a numerical order, separated by commas. This list is followed by a hyphen and then the symbol η. Extended coordination over more than two contiguous carbon atoms should be indicated by an 'en' dash between the locants specifying the extent of the extended coordination, so that, for example, (1–4-η) is preferred to (1,2,3,4-η), and no superscript to the symbol η is then necessary. The locants and the symbol η are enclosed in parentheses.

Examples

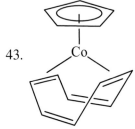

43.

[(1,2,5,6-η)-cyclooctatetraene](η^5-cyclopentadienido)cobalt **or**
[(1,2,5,6-η)-cyclooctatetraene](η^5-cyclopentadienyl)cobalt

44.

dicarbonyl[(1–3-η)-cyclohepta-2,4,6-trien-1-ido](η^5-cyclopentadienido)▶ molybdenum **or**
dicarbonyl[(1–3-η)-cyclohepta-2,4,6-trien-1-yl](η^5-cyclopentadienyl)molybdenum

When unsaturated ligands participate in more than one type of bonding the η symbol may, if necessary, be used in conjunction with the κ symbol. The symbol η then precedes the ligand name while the κ is placed either at the end of the ligand name or, for more complicated structures, after that portion of the ligand name which denotes the particular function in which the ligating atom is found.

Examples

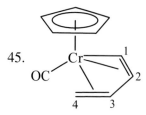

45.

(η^4-buta-1,3-dien-1-ido-κC^1)carbonyl(η^5-cyclopentadienido)chromium **or**
(η^4-buta-1,3-dien-1-yl-κC^1)carbonyl(η^5-cyclopentadienyl)chromium

46.

$[(2–4-\eta)$-but-2-ene-1,1,4-triido-$\kappa C^1]$ carbonyl(η^5-cyclopentadienido)chromium

47.

tricarbonyl[6-oxo-κO-$(2–4-\eta)$-hept-3-en-2-ido]iron(1+) **or**
tricarbonyl[6-oxo-κO-$(2–4-\eta)$-hept-3-en-2-yl]iron(1+)

For more information on the use of kappa (κ) see Section 7.4.5.

Note the use of the arrow in Example 47 to indicate a coordinate (in earlier terminology sometimes also called semi-polar or dative) bond. In the future this kind of formalism, which distinguishes one bond type from another, is likely to be discouraged (see discussion concerning Examples 25 and 26 in Section 8.2.3.3).

If an unsaturated hydrocarbon is a bridging ligand, the prefix μ (see Section 8.2.3.2) may be used in combination with both κ and η. To indicate which atoms of the bridge bind to the various metal atoms, the appropriate locants are cited separated by a colon. If the central atoms of a di- or poly-nuclear structure are the same elements but not equivalently coordinated, in the simplest cases the metal atoms with the higher coordination numbers are assigned the lower locant numbers. For more complex structures, reference 2 should be consulted. The locants of the metal atoms are placed before the η and κ symbols with no hyphens. If ligand locants are also specified, these are separated from the η symbol by a hyphen and the whole expression is enclosed in parentheses, so that a ligand bonding to central atom 1 through contiguous atoms with locants 2, 3, and 4 would be specified as $1(2–4-\eta)$.

Examples

48.

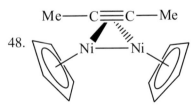

$(\mu$-η^2:η^2-but-2-yne)bis[(η^5-cyclopentadienido)nickel](Ni—Ni) **or**
$(\mu$-η^2:η^2-but-2-yne)bis[(η^5-cyclopentadienyl)nickel](Ni—Ni)

125

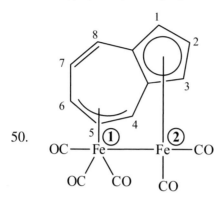

49.

trans-[μ-(1–4-η:5–8-η)-cyclooctatetraene]bis(tricarbonyliron)

50.

{μ-[2(1–3,3a,8a-η):1(4–6-η)]azulene}(pentacarbonyl-1κ³*C*,2κ²*C*)diiron(*Fe—Fe*)

8.2.5.2   *Metallocene nomenclature.* The first transition element sandwich compound containing only carbocycles as ligands was bis(η⁵-cyclopentadienido)iron, [Fe(η⁵-C₅H₅)₂], which has a 'sandwich' structure with two parallel η⁵- or π-bonded rings. It was soon recognised that this compound, like organic aromatic compounds, could undergo electrophilic substitution, which led to the suggestion of the non-systematic name ferrocene, as a parallel with the name benzene, and to similar names for other metallocenes.

Examples

    51. [V(η⁵-C₅H₅)₂]   vanadocene
    52. [Cr(η⁵-C₅H₅)₂]   chromocene
    53. [Co(η⁵-C₅H₅)₂]   cobaltocene
    54. [Ni(η⁵-C₅H₅)₂]   nickelocene
    55. [Ru(η⁵-C₅H₅)₂]   ruthenocene
    56. [Os(η⁵-C₅H₅)₂]   osmocene

Metallocene derivatives may be named using either suffixes to indicate principal groups or prefixes to indicate substituents, in the manner already discussed for substitutive nomenclature in Chapter 6. For metallocene substituent groups 'ocenyl', 'ocenediyl', 'ocenetriyl', *etc.*, are used.

Examples

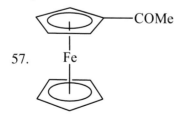

57.

acetylferrocene (using a prefix) **or**
1-ferrocenylethan-1-one (using a suffix)

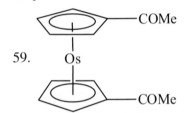

58.

1-[1-(dimethylamino)ethyl]ferrocene (using a prefix) **or**
1-ferrocenyl-*N*,*N*-dimethylethan-1-amine (using a suffix)

Substituents on the equivalent cyclopentadienido rings of the metallocene entity are given the lowest possible numerical locants in the usual manner. The first ring is numbered 1, and if necessary, its individual carbon atoms 1 to 5, and the second ring 1′ and 1′ to 5′.

Examples

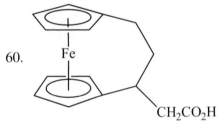

59.

1,1′-diacetylosmocene **or**
1,1′-(osmocene-1,1′-diyl)bis(ethan-1-one)

60.

1,1′-(4-carboxybutane-1,3-diyl)ferrocene **or**
3,5-(ferrocene-1,1′-diyl)pentanoic acid

61. $[Ru(\eta^5\text{-}C_5Me_5)_2]$
decamethylruthenocene **or**
bis(pentamethyl-η^5-cyclopentadienido)ruthenium **or**
bis(pentamethyl-η^5-cyclopentadienyl)ruthenium

62. $[Cr(\eta^5\text{-}C_5Me_4Et)_2]$
1,1′-diethyloctamethylchromocene **or**
bis(1-ethyl-2,3,4,5-tetramethyl-η^5-cyclopentadienido)chromium **or**
bis(1-ethyl-2,3,4,5-tetramethyl-η^5-cyclopentadienyl)chromium

Oxidized species such as $[Fe(\eta^5\text{-}C_5H_5)_2]^+$ are often referred to as metallocenium($n+$) salts, though the suffix 'ium' does not carry the usual meaning it has in substitutive nomenclature, *i.e.*, the addition of a hydron to a neutral parent compound. Consequently the systematic additive names for such

127

CHAPTER 8

species are strongly preferred. For instance, $[Fe(\eta^5\text{-}C_5H_5)_2]^+$ should be named bis(η^5-cyclopentadienido)iron(1+), rather than ferrocenium(1+). The same applies to substituted derivatives.

Example
63. $[Co(\eta^5\text{-}C_5H_5)(\eta^5\text{-}C_5H_4COMe)][PF_6]$
(acetyl-η^5-cyclopentadienido)(η^5-cyclopentadienido)cobalt(1+) hexafluoridophosphate(1–) **or**
(acetyl-η^5-cyclopentadienyl)(η^5-cyclopentadienyl)cobalt(1+) hexafluoridophosphate(1–)

Metallocene names should not be applied to all superficially similar derivatives of all transition elements. For example, manganocene is a misnomer since $[Mn(\eta^5\text{-}C_5H_5)_2]$ has a chain structure in the solid state, with no individual sandwich entities involved. However, decamethylmanganocene, $[Mn(\eta^5\text{-}C_5Me_5)_2]$, has a normal sandwich structure, as does decamethylrhenocene, $[Re(\eta^5\text{-}C_5Me_5)_2]$. The 'ocene' names should therefore be confined to discrete molecules where the cyclopentadienido rings are essentially parallel, and the metal is in the d-block of the periodic table. The terminology should not be applied to compounds of the s- or p-block elements such as $Ba(C_5H_5)_2$ or $Sn(C_5H_5)_2$.

Many compounds have ligands additional to the two η^5-cyclopentadienido rings, and they are often named as metallocene di(ligand) species. For example, $[Ti(\eta^5\text{-}C_5H_5)_2Cl_2]$ is frequently named titanocene dichloride or dichlori▶ dotitanocene. This practice is discouraged since metallocene nomenclature applies strictly only to compounds in which the two rings are parallel. Thus, dichlori▶ dobis(η^5-cyclopentadienido)titanium is the correct name for $[Ti(\eta^5\text{-}C_5H_5)_2Cl_2]$. The correct name for $[Zr(\eta^5\text{-}C_5H_5)_2Me_2]$ is not dimethylzirconocene but bis(η^5-cyclopentadienido)dimethanidozirconium.

8.3   ORGANOMETALLIC COMPOUNDS OF THE MAIN
        GROUP ELEMENTS

8.3.1   **General**

In principle, all organometallic compounds, whether of the Transition Group or of the Main Group elements, may be given additive (coordination) names provided the constitution of the compound is known. However, in addition to compounds of the metals and semi-metals, many compounds of elements such as boron, silicon, arsenic and selenium are often considered to be organometallic, and compounds of these elements are commonly named by notionally substituting the hydrogen atoms of a parent hydride by the appropriate substituent groups.

The current recommendation (reference 2) is that organometallic compounds derived from the elements of Groups 13–16 be named by substitutive nomenclature, whereas those derived from the elements of Groups 1 and 2 be named

128

using additive nomenclature or, in some cases, simply compositional nomenclature if less structural information is to be conveyed. Where an organometallic compound contains two or more possible central atoms (which may be associated with different nomenclature systems according to the above recommendation), a choice of the basis of the name must be made. Such polynuclear organometallic compounds are beyond the scope of this text, but guidance for their naming may be found in Section IR-10.4 of reference 2.

8.3.2 **Organometallic compounds of Groups 1 and 2**

Organometallic compounds of defined structure of the elements of Groups 1 and 2 are named using the additive nomenclature, as described in Chapter 7. Thus, prefixes denoting the organic groups and any other ligands are placed in alphabetical order before the name of the metal. As before, these prefixes adopt the additive 'ido', 'diido', *etc.* suffixes, but the substitutive 'yl', 'diyl', *etc.* suffixes will often be encountered in the literature (see Sections 8.2.2, 8.2.3 and 8.2.4). The presence of a hydrogen atom attached to the metal centre must always be indicated by the prefix 'hydrido' (Example 1). The name of a cyclic compound with the central atom in the ring may be formed using appropriate locants of a divalent 'diido' group (or, alternatively, a 'diyl' group) to indicate chelate-type bonding to the metal, as in Example 2.

Examples

1. [Be(Et)H]   ethanidohydridoberyllium **or** ethyl(hydrido)beryllium

2.

$[2\text{-}(4\text{-methylpent-3-en-1-yl})\text{but-2-ene-1,4-diido-}\kappa^2 C^1,C^4]$magnesium **or**
$[2\text{-}(4\text{-methylpent-3-en-1-yl})\text{but-2-ene-1,4-diyl}]$magnesium

Many organometallic compounds of Groups 1 and 2 exist in associated molecular forms (as aggregates) or contain structural solvent, or both. Often complex equilibria exist between the different species of an organolithium reagent or a Grignard reagent in solution, and these are markedly dependent on temperature, concentration and solvent. Consequently, and to be practical, their names are usually based solely on the stoichiometric compositions of the compounds, unless it is specifically desired to draw attention to the extent of aggregation or the nature of any structural solvent, or both (see Example 3 below). In Examples 4 and 5, the different types of name for what is essentially the same material reflect the different structures implied by the formulae. As usual, the formulae enclosed in square brackets designate coordination entities.

Examples

   3. LiPh lithium benzenide (*compositional name*)

      [{Li(OEt$_2$)(μ_3-Ph)}$_4$]
      tetrakis[(μ_3-benzenido)(ethoxyethane)lithium]

   4. LiMe   lithium methanide (*compositional name*)

      (LiMe)$_n$ poly(methanidolithium) **or** poly(methyllithium)

      [LiMe]   methanidolithium **or** methyllithium

      [(LiMe)$_4$]   tetra-μ_3-methanido-tetralithium **or**
                        tetra-μ_3-methyl-tetralithium

   5. MgIMe   magnesium iodide methanide (*simple compositional name*)
      [MgMe]I   methanidomagnesium iodide **or** methylmagnesium iodide
                        (*compositional name with formally electropositive component
                        named by additive nomenclature*)

      [MgI(Me)] iodido(methanido)magnesium **or** iodido(methyl)magnesium
                        (*additive name of coordination type*)

      [MgI(Me)]$_n$ poly[iodido(methanido)magnesium] **or**
                        poly[iodido(methyl)magnesium]

Metallocene terminology (Section 8.2.5.2) is not recommended for bis(cyclo-pentadienido) compounds of the Main Group metals. Thus, [Mg(η^5-C$_5$H$_5$)$_2$] should be named bis(η^5-cyclopentadienido)magnesium.

8.3.3   **Organometallic compounds of Groups 13–16**

Organometallic compounds of the elements of Groups 13–16 are named using sub-stitutive nomenclature (Chapter 6). The name of the parent hydride (Section 6.2.10) is modified by a prefix for each substituent replacing a hydrogen atom of the parent hydride. The prefix should be in appropriate substituent form (chloro, methyl, sulfanylidene, *etc.*) and not in ligand form (chlorido, methanido, sulfido, *etc.*).

Where there is more than one kind of substituent, the prefixes are cited in alphabetical order before the name of the parent hydride, parentheses being used to avoid ambiguity, and multiplicative prefixes being used as necessary. Non-standard bonding numbers are indicated using the λ-convention (see Section 6.2.10). An overview of the rules for naming substituted derivatives of parent hydrides is given in Section 6.2, and a detailed exposition may be found in reference 1.

Examples

   6. AlH$_2$Me   methylalumane
   7. AlEt$_3$ triethylalumane
   8. Me$_2$CHCH$_2$CH$_2$In(H)CH$_2$CH$_2$CHMe$_2$
                            bis(3-methylbutyl)indigane
   9. Sb(CH=CH$_2$)$_3$ triethenylstibane **or** trivinylstibane
  10. SbMe$_5$ pentamethyl-λ^5-stibane
  11. PhSb=SbPh   diphenyldistibene

12. $GeCl_2Me_2$ dichlorodimethylgermane

13. $GeMe(SMe)_3$ methyltris(methylsulfanyl)germane

14. BiI_2Ph diiodo(phenyl)bismuthane

15. $Et_3PbPbEt_3$ hexaethyldiplumbane

16. $SnMe_2$ dimethyl-λ^2-stannane

17. $BrSnH_2SnCl_2SnH_2(CH_2CH_2CH_3)$

                             1-bromo-2,2-dichloro-3-propyltristannane

18. $Me_3SnCH_2CH_2C{\equiv}CSnMe_3$ (but-1-yne-1,4-diyl)bis(trimethylstannane)

Characteristic groups (for example, $-NH_2$, $-OH$, $-COOH$, *etc.*) may be indicated using suffixes. The name of the parent hydride carrying the highest-ranking such group is modified by the suffix, and other substituents are then denoted by prefixes as described in Section 6.6. If the species to be named is a substituent group, the name of the corresponding Group 13–16 parent hydride is modified by changing the ending 'ane' to 'anyl' (or 'yl' for the Group 14 elements), 'anediyl', *etc.*

Examples

19. $(EtO)_3GeCH_2CH_2COOMe$

     methyl 3-(triethoxygermyl)propanoate

20. $H_2As[CH_2]_4SO_2Cl$

     4-arsanylbutane-1-sulfonyl chloride

21. $OCHCH_2CH_2GeMe_2GeMe_2CH_2CH_2CHO$

     3,3'-(1,1,2,2-tetramethyldigermane-1,2-diyl)dipropanal

22. $SiMe_3NH_2$

     trimethylsilanamine

It may be necessary or preferable to consider a parent hydride in which several (four or more) skeletal carbon atoms of an unbranched hydrocarbon have been replaced by Main Group elements. In this method of **skeletal replacement** the heteroatoms are designated by the 'a' terms of replacement nomenclature (see Table 6.2) and cited in the element sequence given by Table P3 and preceded by the appropriate locant(s). The locants are assigned by numbering the chain from that end which gives lower locants to the heteroatom set as a whole and, if these are equal, from that end which gives the lower locant or locant set to the replacement prefix first cited. If there is still a choice, lower locants are assigned to the sites of unsaturation. This nomenclature is fully described in reference 1.

Examples

23. $\overset{2}{}Me\overset{3}{Si}H_2\overset{4}{C}H_2\overset{5}{C}H_2\overset{6}{Si}H_2\overset{7}{C}H_2\overset{8}{C}H_2\overset{9}{Si}H_2\overset{10}{C}H_2\overset{11}{C}H_2\overset{}{Si}H_2Me$

     2,5,8,11-tetrasiladodecane

24. $Me\overset{2}{Si}H_2\overset{3}{O}\overset{4}{P}\overset{5}{H}OCH_2Me$

     3,5-dioxa-4-phospha-2-silaheptane

25. $\overset{1}{H}S\overset{2}{C}H{=}\overset{3}{N}\overset{4}{O}CH_2\overset{5}{S}e\overset{6}{C}H_2\overset{7}{O}\overset{8}{N}HMe$

     3,7-dioxa-5-selena-2,8-diazanon-1-ene-1-thiol

When elements from Groups 13–16 replace carbon atoms in mono-cyclic systems, the resulting structures may be named using the extended Hantzsch–Widman procedures. This nomenclature is described in Section 6.3.3 and in reference 1 and will not be elaborated further here. A more comprehensive treatment of the nomenclature of organic compounds containing the elements of Groups 13–16 may be found in reference 1.

REFERENCES

1. *Nomenclature of Organic Chemistry, IUPAC Recommendations*, eds. W. H. Powell and H. Favre, RSC Publishing, Cambridge, U.K., in preparation (as of October 2 2011). Until then the current publications should be consulted. These are: *Nomenclature of Organic Chemistry, 1979*; *A Guide to IUPAC Nomenclature of Organic Compounds, Recommendations 1993*; and any information on the subject at issue available at http://www.iupac.org. New recommendations are published in instalments on the Web by IUPAC as soon as they are finally approved by the usual IUPAC review procedures.
2. *Nomenclature of Inorganic Chemistry, IUPAC Recommendations 2005*, eds. N. G. Connelly, T. Damhus, R. M. Hartshorn and A. T. Hutton, RSC Publishing, Cambridge, UK.
3. F. A. Cotton, *Journal of the American Chemical Society*, 1968, **90**, 6230–6232.

9 Nomenclature of Polymers

INTRODUCTION

Reference 1 contains a glossary of the terms generally used in polymer chemistry (reference 2), and this should be used whenever clarification of definitions is required. It is of fundamental importance to understand the distinction between the terms 'polymer' and 'macromolecule'. A **polymer** is a substance composed of **macromolecules** (or polymer molecules) of different molar masses with an average value usually in excess of $10^4 \, g \, mol^{-1}$. Each macromolecule comprises a sequence or array of constitutional units connected by covalent bonds. A polymer is named using the names of these constitutional units, or portions or multiples thereof, preceded by the prefix 'poly'. A macromolecule is generally so large that the addition or removal of a single repeating unit does not affect significantly its molecular properties. However, polymers usually do not possess a uniform structure but consist of a mixture of individual macromolecules which may differ in chain length and/or structural arrangement. The latter can include irregularities in the sequence of constitutional units, constitutional-unit orientation, branching, and end-group structure. Thus a polymer differs in a number of ways from common, low-molecular-weight (non-polymeric) substances.

There are two principal approaches to naming polymers (reference 3), **source-based nomenclature** and **structure-based nomenclature**. The former uses the names of the monomer or monomers from which the polymer is presumed to have been prepared and the latter uses, where possible, IUPAC nomenclature for the constitutional units of the constituent macromolecules. The two approaches are exemplified by the following example. The polymer prepared by polymerisation of the monomer propene may be called polypropene. This is the source-based name. The structure-based name is derived from the name of the constitutional unit as it occurs within the macromolecule. In this case, the structure is $-CH(CH_3)CH_2-$, and the polymer is therefore called poly(1-methylethane-1,2-diyl).

In source-based nomenclature, when the name of the monomer comprises more than one word, it is enclosed in parentheses, as in poly(methyl methacrylate), but when the monomer name is a single word parentheses are not necessary unless ambiguity might otherwise arise. Parentheses are always used in structure-based nomenclature.

For a linear homopolymer and a copolymer of regular repeating structure, the smallest unit, the repetition of which gives the structure of its constituent macromolecules, is known as the **constitutional repeating unit** (**CRU**) (reference 2). The CRU forms the basis of the structure-based name and the structural representation of the polymer. Thus, the structure-based name of polypropene, poly(1-methylethane-1,2-diyl), is of the form poly(CRU) and the structural

Principles of Chemical Nomenclature: A Guide to IUPAC Recommendations
Edited by G. J. Leigh
© International Union of Pure and Applied Chemistry 2011
Published by the Royal Society of Chemistry, www.rsc.org

representation is $-(CH(CH_3)CH_2-)_n$, where n indicates the polymeric nature of the molecules. For such regular polymers, both structure-based and source-based nomenclatures are allowed IUPAC nomenclatures (reference 3). However, for most copolymers (reference 4) and for non-linear macromolecules and their assemblies (reference 5), molecular structures are not known precisely and therefore source-based names are used.

In addition to structure-based and source-based names, established, traditional names are also allowed by IUPAC for some frequently encountered polymers, for example, polyethylene, $-(CH_2-)_n$, polypropylene, $-(CH(CH_3)CH_2-)_n$, and polystyrene, $-(CH(C_6H_5)CH_2-)_n$. The structure-based names of polyethylene, polypropylene and polystyrene are poly(methylene), poly(1-methylethylene) or poly(1-methylethane-1,2-diyl), and poly(1-phenylethane-1,2-diyl), respectively. Some abbreviations recommended by the International Organization for Standardization (ISO) are also allowed, provided they are fully defined when they are first used in a text. For example, PE, PP, and PS are the allowed abbreviations for polyethylene, polypropylene, and polystyrene.

Some selected definitions of important terms in polymer science are given below (references 1, 2 and 3).

- A **polymer** is a substance composed of macromolecules.
- A **macromolecule**, or **polymer molecule**, is a molecule of high relative molecular mass, the structure of which essentially comprises the multiple repetition of units derived from molecules of low relative molecular mass.
- A **monomer molecule** is a molecule that can undergo polymerisation, thereby contributing constitutional units to the essential structure of a macromolecule.
- A **monomer** is a substance composed of monomer molecules that can undergo polymerisation, thereby contributing constitutional units to the essential structure of a macromolecule.
- An **oligomer molecule** is a molecule of intermediate molecular mass, the structure of which essentially comprises a small number of units derived, actually or conceptually, from molecules of lower relative mass.
- An **oligomer** is a substance composed of oligomer molecules.
- A **constitutional unit** is an atom or group of atoms (with pendant atoms or groups, if any) comprising a part of the essential structure of a macromolecule.
- The process of **polymerisation** is the conversion of a monomer, or of a mixture of monomers, into a polymer.
- A **monomeric unit** is the largest constitutional unit contributed by a single monomer molecule to the structure of a macromolecule.
- A **constitutional repeating unit** (CRU) is the smallest constitutional unit, the repetition of which constitutes a regular macromolecule.
- A **regular polymer** is a polymer composed of regular macromolecules.
- A **regular macromolecule** is a macromolecule, the structure of which essentially comprises the repetition of a single constitutional unit with all units connected identically with respect to directional sense.
- An **irregular polymer** is a polymer composed of irregular macromolecules.
- An **irregular macromolecule** is a molecule, the structure of which essentially comprises the repetition of more than one type of constitutional unit, or of macromolecules, the structure of which comprises constitutional units not all connected identically with respect to directional sense.

- A **homopolymer** is a polymer derived from one species of monomer.
- A **copolymer** is a polymer derived from more than one species of monomer.
- A **single-strand polymer** is a polymer composed of single-strand macromolecules.
- A **single-strand macromolecule** is a macromolecule, the structure of which comprises constitutional units connected in such a way that adjacent constitutional units are joined to each other through two atoms, one on each constitutional unit.
- A **double-strand polymer** is a polymer composed of double-strand macromolecules.
- A **double-strand macromolecule** is a macromolecule, the structure of which comprises constitutional units connected in such a way that adjacent constitutional units are joined to each other through three or four atoms, two on one side of each constitutional unit and one or two on the other. Examples of a double-strand macromolecule are a **ladder macromolecule**, which consists of an uninterrupted sequence of rings with adjacent rings having two or more atoms in common, and a **spiro macromolecule**, which consists of an uninterrupted sequence of rings, with adjacent rings having only one atom in common.
- A **block** is a portion of a macromolecule, comprising many constitutional units, that has at least one constitutional or configurational feature that is not present in adjacent portions.
- A **block copolymer** is a polymer composed of block macromolecules.
- A **block macromolecule** is a macromolecule composed of blocks in linear sequence.
- A **graft copolymer** is a polymer composed of graft macromolecules.
- A **graft macromolecule** is a macromolecule with one or more species of block connected to the main chain as side-chains, these side-chains having constitutional or configurational features that differ from those in the main chain (see reference 2 for definitions of various types of chain).
- An **end-group** is a constitutional unit that is an extremity of a macromolecule.

Problems with determining the appropriate nomenclature for macromolecules arise not so much from the size and molar mass distribution (dispersity) but much more from their lack of structural identity. Structure representations are generally idealised, so that names of polymers indicate an almost perfect regularity and uniformity. The frequently undefined end-groups are usually ignored for nomenclature purposes. However, other significant but indefinite deviations from ideality often lend themselves only to qualitative statements, such as that a certain stereoregularity is 'high'. This lack of rigour is inevitable.

Of the two possible approaches to naming polymers, structure-based and source-based (Section 9.3), the former is preferred in IUPAC nomenclature. Structure-based nomenclature is described in Section 9.2 below and pays no regard to the polymer's origin. Name construction starts from a consideration of the structure (chemical constitution) of the macromolecules of which the polymer is comprised. For a structure-based name to be formulated with confidence, a detailed knowledge of the structure of the macromolecule and its analysis are required. However, except for **homopolymers**, which are derived from only one monomer, and simple **copolymers**, which are derived from more than one species of monomer, the detailed stucture upon which this analysis can be based is rarely available. In addition, although structure-based nomenclature is strictly systematic and results in an unambiguous name for any given polymer of known structure, this name is often too complex or too long for ready use.

In contrast, the source-based nomenclature described in Section 9.3 allows for the construction of a name for a polymer from the name of the monomer from which it was derived. Such names are usually simpler than the structure-based names and are more easily understood, but they may also have their drawbacks. There are cases where, depending on reaction conditions, the polymerisation of a given monomer can result in polymers of different structure (*cf.,* Section 9.4), and deciphering a source-based name frequently requires some knowledge of organic and inorganic chemistry.

The simplest polymers consist of linear-chain macromolecules. However, some polymers have branches of different types, some contain rings, and individual chains can be linked or cross-linked to form networks of extensive size. Different networks can interpenetrate. Consideration of these structures is outside the scope of this introductory text, but they are discussed in the recently published second edition of the so-called Purple Book, the *Compendium of Polymer Terminology and Nomenclature* (reference 1). Macrocyclic structures are covered by the 2008 IUPAC recommendations (reference 6).

9.2 STRUCTURE-BASED NOMENCLATURE (references 1 and 6–8)

9.2.1 **Regular single-strand organic polymers** (reference 7)

The most common species likely to be met in an elementary text is a regular organic polymer. The structure of its macromolecules essentially comprises the repetition of only one constitutional repeating unit (CRU) connected identically with respect to direction. In a **single-strand** macromolecule adjacent constitutional units are joined to each other through two atoms, one on each constitutional unit. The name of such a polymer takes the form poly(CRU), because structure-based names always have the CRU in enclosing marks after the prefix 'poly'. If multiple sets of enclosing marks are needed, IUPAC recommends the use of curved brackets (parentheses) for the innermost application, then square brackets, then curly brackets, so the nesting order of enclosing marks in polymer nomenclature is $\{[()]\}$.

A regular single-strand polymer may be represented by a formula such as shown in Example 1, and the various parts of this formula are defined as shown.

Example

1. $E^1\!\!-\!\!(A\!-\!B\!-\!C\!-\!D)_n\!\!-\!E^2$ regular single-strand polymer
   $\qquad R^1 \qquad R^2$

   $-A\!-\!B\!-\!C\!-\!D-$ constitutional repeating unit (CRU)
   $\quad R^1 \qquad R^2$

   $-A-\,,\ -B-\,,\ -C-\,,\ -D-$ subunits

   $-B-\,,\ -D-$ substituted subunits
   $\quad R^1 \qquad R^2$

   R^1, R^2 substituents to subunits

   E^1, E^2 end-groups

There is often more than one way to write a CRU for a macromolecule. In simple cases such as that shown in Example 2 these are readily identified.

Example

2. 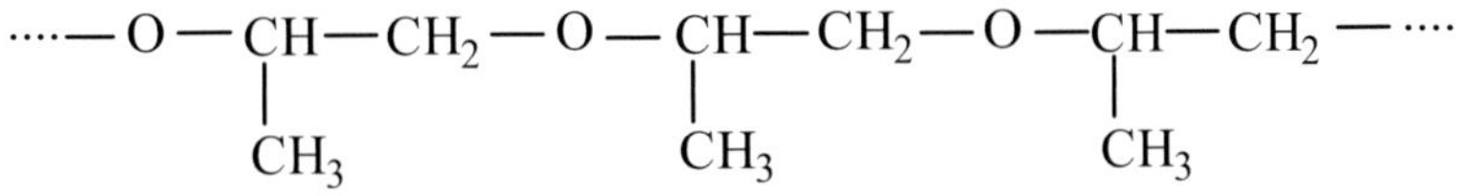 ·····—O—CH—CH$_2$—O—CH—CH$_2$—O—CH—CH$_2$—·····
            | | |
            CH$_3$ CH$_3$ CH$_3$

For this chain one might select one of three possible CRUs, each with two possible orientations, which makes six possibilities in all. Each may be divided into an oxygen-atom subunit and a carbon chain.

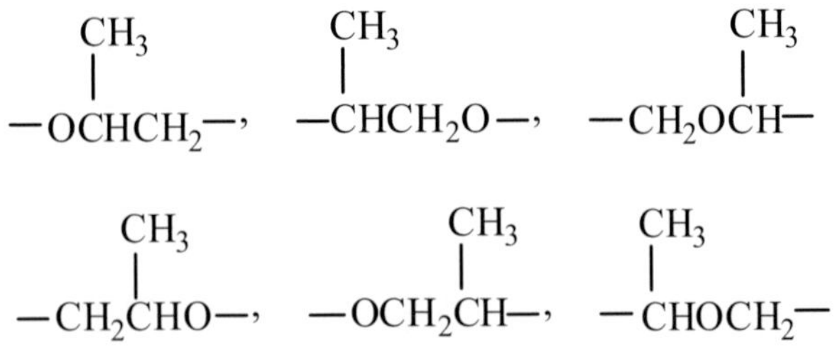

      CH$_3$ CH$_3$ CH$_3$
      | | |
 —OCHCH$_2$—,   —CHCH$_2$O—,   —CH$_2$OCH—

        CH$_3$ CH$_3$ CH$_3$
        | | |
 —CH$_2$CHO—,   —OCH$_2$CH—,   —CHOCH$_2$—

The preferred CRU consists of one or more divalent organic groups (subunits), named according to the nomenclature rules of organic chemistry, see Chapters 4, 5, and 6 of this book, and also Table P9 and reference 7.

A set of rules has been developed to specify the seniority among subunits, in order to identify the atom or group where the preferred CRU starts and also the direction in which to move along the chain to reach the end of the preferred CRU (orientation). The order of seniority of subunits is of primary importance in the generation of polymer names.

The basic order of seniority among the types of divalent groups is:

heterocyclic systems > heteroatom chains > carbocyclic systems > carbon chains.

Example 2, above, may be used to illustrate the application of these criteria.
1. Oxygen is the senior subunit.
2. For the substituted ethylene (= ethane-1,2-diyl) subunit, there is a choice of numbering to indicate the position of the methyl group. The CRU may be written –OCH(CH$_3$)CH$_2$– or –OCH$_2$CH(CH$_3$)–. The normal practice is to write the CRU from left to right, with the chain atoms numbered consecutively in the same direction. The preferred position for the methyl substituent is on the leftmost carbon atom, *i.e.*, it is assigned the lowest possible locant when reading from left to right. The name of the polymer would thus appear to be one of two possibilities:

poly[oxy(1-methylethylene)]
or
poly[oxy(1-methylethane-1, 2- diyl)]

These names correspond to a polymer entirely composed of –OCH(CH$_3$)CH$_2$– units. Further examples of important polymers are given in Table 9.1.

CHAPTER 9

Table 9.1 Structure-based and source-based names for commonly encountered polymers (adapted from reference 1)

Structure	Source-based name	Structure-based name
$+CH_2+_n^{\,a}$	polyethene polyethylene[b,c]	poly(methylene)
$+CHCH_2+_n$ | CH_3	polypropene polypropylene[b]	poly(1-methylethane-1,2-diyl) poly(1-methylethylene)
CH_3 | $+CCH_2+_n$ | CH_3	poly(2-methylpropene) polyisobutylene[b] polyisobutene	poly(1,1-dimethylethane-1,2-diyl)
$+CH=CH-CH_2CH_2+_n$	poly(buta-1,3-diene) polybutadiene[b]	poly(but-1-ene-1,4-diyl)
$+CH_2C=CH-CH_2+_n$ | CH_3	polyisoprene[b]	poly(1-methylbut-1-ene-1,4-diyl)
$+CHCH_2+_n$ | C_6H_5	poly(ethenylbenzene) polystyrene[b]	poly(1-phenylethane-1,2-diyl) poly(1-phenylethylene)
$+CHCH_2+_n$ | CN	polyacrylonitrile[b]	poly(1-cyanoethane-1,2-diyl)
$+CHCH_2+_n$ | OH	poly(vinyl alcohol)	poly(1-hydroxyethane-1,2-diyl)
$+CHCH_2+_n$ | $OCOCH_3$	poly(vinyl acetate)	poly(1-acetoxyethane-1,2-diyl)
$+CHCH_2+_n$ | Cl	poly(vinyl chloride)	poly(1-chloroethane-1,2-diyl)
$+CF_2CH_2+_n$	poly(1,1-difluoroethene) poly(vinylidene difluoride)[b]	poly(1,1-difluoroethane-1,2-diyl)
$+CF_2+_n^{\,a}$	poly(tetrafluoroethene) poly(tetrafluoroethylene)[b]	poly(difluoromethylene)
(cyclic acetal structure)$-CH_2-_n$ $O\ \ O$ $CH_2CH_2CH_3$	poly(vinyl butyral) poly(butyraldehyde divinyl acetal)	poly[(2-propyl-1,3-dioxane-4,6-diyl)methylene]
$+CHCH_2+_n$ | $COOCH_3$	poly(methyl acrylate)	poly[1-(methoxycarbonyl)ethane-1,2-diyl]
CH_3 | $+CCH_2+_n$ | $COOCH_3$	poly(methyl methacrylate)	poly[1-(methoxycarbonyl)-1-methylethane-1,2-▶diyl]
$+OCH_2+_n$	polyformaldehyde	poly(oxymethylene)
$+OCH_2CH_2+_n$	poly(ethylene oxide)[b]	poly(oxyethane-1,2-diyl)
$+O-\langle\bigcirc\rangle+_n$	poly(1,4-phenylene oxide)	poly(oxy-1,4-phenylene)
$+(CH_2)_2-O-\overset{O}{\underset{\|\|}{C}}-\langle\bigcirc\rangle-\overset{O}{\underset{\|\|}{C}}-O+_n$	poly(ethane-1,2-diyl terephthalate) poly(ethylene terephthalate)	poly(oxyethane-1,2-diyloxyterephthaloyl)
$+NHCO(CH_2)_5+_n$	poly(hexano-6-lactam) poly(ε-caprolactam)[b]	poly[imino(1-oxohexane-1,6-diyl)]
$+NHCH_2CH_2+_n$	poly(aziridine) poly(ethylenimine)[b]	poly(iminoethane-1,2-diyl)

[a] the formulae $-(CH_2CH_2)_n-$ and $-(CF_2CF_2)_n-$ are more often used; they are acceptable due to past usage and in an attempt to retain some similarity to the formulae of the CRUs of homopolymers derived from other ethene derivatives.

[b] traditional names, some of which are based upon monomer names which are no longer acceptable.

[c] The name ethylene should only be used for the divalent group $-CH_2-CH_2-$, and not for the monomer $CH_2=CH_2$, which is named ethene.

This example also shows that within each type of CRU rules of seniority are applied, which are based on the nature of the subunits (kind, size or both). Among identical types of constitutional unit, the rules are based upon their degree of unsaturation and substituents (number, kind and locants). The rules are applied consecutively until a decision is reached. Further details for a more advanced treament have been published elsewhere (references 7 and 9).

In polymers containing rings and ring systems in the polymer chain, the rules of the nomenclature of organic chemistry (reference 9) for the seniority of heterocyclic systems apply. For carbocyclic systems the major critera are ring size and degree of unsaturation. For these rules see Chapter 6. They apply also to acyclic polymer chains, as presented in reference 7.

If the end-groups of a polymer chain are known, they may be specified by affixing prefixes to the name of the polymer; the symbols α and ω being used to designate the left-hand and right-hand end-groups, respectively, as in Example 3.

Example

3.

α-(trichloromethyl)-ω-chloropoly(1,4-phenylenemethylene)

If a regular polymer chain is attached to another polymer chain or to an organic low-molecular-weight structure, it is treated as substituent to the chain or structure. For details, see reference 7.

9.2.2 **Structure-based Nomenclature for Irregular Single-Strand Organic Polymers** (reference 8)

Polymers sometimes consist of chains in which the repetition of the structure is not regular but irregular. The macromolecules of **irregular polymers** are composed of constitutional units not all connected identically with respect to direction, or consist of more than one type of constitutional unit. As for regular single-strand polymers, irregular polymers are named by placing the prefix 'poly' before the structure-based names of the constitutional units, which are enclosed in parentheses, the names of individual constitutional units being separated by oblique strokes. Examples 4–6 illustrate some typical polymers described schematically by the formula poly(A/B).

Examples

4. A partially hydrolysed head-to-tail poly(vinyl acetate) may contain the constitutional units shown.

poly(1-acetoxyethane-1,2-diyl/1-hydroxyethane-1,2-diyl)

5. A copolymer of vinyl chloride and styrene with a head-to-tail arrangement of units.

$$-\overset{\displaystyle |}{\underset{\displaystyle Cl}{CH}}-CH_2- \quad \text{and} \quad -CH-CH_2-$$

poly(1-chloroethane-1,2-diyl/1-phenylethane-1,2-diyl)

6. A chlorinated polyethylene consisting of three constitutional units.

$$-\overset{\displaystyle |}{\underset{\displaystyle Cl}{CH}}- \quad -\overset{\displaystyle Cl}{\underset{\displaystyle Cl}{\overset{|}{\underset{|}{C}}}}- \quad -CH_2-$$

poly(chloromethylene/dichloromethylene/methylene)

A **block copolymer** is composed of block macromolecules in which adjacent segments, called blocks, are comprised of different CRUs. The blocks may be derived from different monomers or from the same species of monomer but with different compositions or sequence distributions of constitutional units. Block copolymers are named by sequentially citing the blocks, typically as poly(A)–poly(B)–poly(C). A block copolymer, in which the blocks are joined by a specific junction unit, can be named as in Example 7.

Example

7. Blocks of poly(ethylene oxide) and poly(vinyl chloride) joined by a dimethylsilanediyl unit, $-Si(CH_3)_2-$. Dashes, not hyphens, are used to indicate the chemical bonds between the blocks and junction units.

$$\left(CH_2-CH_2-O\right)_n , \quad -\overset{\displaystyle CH_3}{\underset{\displaystyle CH_3}{\overset{|}{\underset{|}{Si}}}}- , \quad \left(\overset{\displaystyle |}{\underset{\displaystyle Cl}{CH}}-CH_2\right)_n$$

poly(ethane-1,2-diyloxy)-dimethylsilanediyl-poly(1-chloroethane-1,2-diyl)

9.3   SOURCE-BASED NOMENCLATURE (references 4 and 5)

For some polymers a structure-based name will be very complicated or long or – because of lack of sufficient structural information – even impossible to construct. In such cases source-based nomenclature is the method of choice. For simple homopolymers such as polystyrene and poly(vinyl acetate), source-based nomenclature continues to be used because of its simplicity, convenience and obvious relationship to the monomers from which the polymers are prepared, even though structure-based names may have become available for many polymers and, in principle, are more precise.

The source-based name of the polymer is formed by attaching the prefix 'poly' to the name of the real or presumed monomer, or starting material (source), from which the polymer is derived. When the name of the source consists of more than one word, or any ambiguity is anticipated, the name of the source is parenthesised.

Examples
  1. polyacrylonitrile
  2. poly(methyl methacrylate)
  3. polystyrene
  4. poly(vinyl acetate)
  5. poly(vinyl chloride)

For copolymers, an italicised **connective** or **infix** is inserted, to convey what is known about the arrangement of the constitutional units.

Examples
  6. *-co-* for an unknown or unspecified arrangement:
      poly[styrene-*co*-(methyl methacrylate)].
  7. *-stat-* for a statistical arrangement, obeying known statistical laws:
      poly(styrene-*stat*-acrylonitrile-*stat*-buta-1,3-diene).
  8. *-ran-* for a random arrangement, with a Bernoullian distribution:
      poly[ethene-*ran*-(vinyl acetate)].
  9. *-alt-* for an alternating sequence:
      poly[(ethylene glycol)-*alt*-(terephthalic acid)] and also for an unspecified arrangement of two alternating pairs of monomers: poly{[(ethylene glycol)-*alt*-(terephthalic acid)]-*co*-[(ethylene glycol)-*alt*-(isophthalic acid)]}.
  10. *-block-* for linear arrangements of blocks such as -AAAA-BBB-:
      polystyrene-*block*-poly(buta-1,3-diene)
  11. *-graft-* for a graft arrangement:poly(buta-1,3-diene)-*graft*-polystyrene.

Such a graft arrangement is depicted schematically Example 12.

Example

  12.   polyA-*graft*-polyB

           —AAAA'AAA—
                  |
                  B
                  B
                  B
                  |

More complex structures such as branched, comb, cyclic, graft, network, and star polymers are covered by the 1997 IUPAC recommendations (references 5, 6 and 8).

9.4   GENERIC SOURCE-BASED NOMENCLATURE (reference 10)

A given polymer may have more than one source-based name, for example, when it can be prepared in more than one way. If on the other hand a monomer or a pair of complementary monomers can give rise to more than one polymer, or if the polymer is obtained through a series of intermediate structures, simple source-based nomenclature would be inadequate because a simple source-based name would be identical for the different products. For example the source-based name poly(buta-1,3-diene) does not indicate whether the CRUs in the polymer are the 1,2-, 1,4-*cis*-, or 1,4-*trans*-isomers. The ambiguity is easily resolved by incorporating the class name (generic name) of the corresponding polymer which describes the most appropriate type of characteristic group, aromatic rings, cycloalkanes or heterocyclic system, *etc.* This is helpful even in cases where there is only one product formed, in particular when the structure is complex.

In the generic source-based nomenclature system, all the published rules on source-based nomenclature (references 1, 4 and 5) can be applied (see Section 9.3), with the addition of the generic part of the name. In such a name the polymer class (generic) name (polyG) is followed by a colon and then by the actual or hypothetical monomer name(s) (A, B, *etc.*). These are parenthesised in the case of a copolymer (Examples 3 and 4) and for reasons of clarity often also for homopolymers (Example 2).

Examples
1. polyG:A
2. polyG:(B)
3. polyG:(A-*co*-B)
4. polyG:(A-*alt*-B)

Example 5 illustrates the application of generic source-based nomenclature in a specific case.

5. $\left[\text{CH–CH}_2\right]_n$
   $\quad$ CH=CH$_2$

The source-based name is poly(buta-1,3-diene) the structure-based name is poly(1-vinylethane-1,2-diyl), and the generic source-based name is polyalkylene:(buta-1,3-diene). Note that there is the possiblity of an isomeric form for the CRU, Example 6, giving rise to the structure-based name poly(but-1-ene-1,4-diyl), but with the same source-based name as Example 5. However, the generic source-based names allow a distinction to be made between the two forms, polyalkylene:(buta-1,3-diene), Example 5, and polyalkenylene:(buta-1,3-diene), Example 6.

6. $\left[\text{CH=CH–CH}_2\text{–CH}_2\right]_n$

9.5 NOMENCLATURE OF REGULAR DOUBLE-STRAND ORGANIC POLYMERS (reference 11)

A double-strand polymer is a polymer, the molecules of which are formed by an uninterrupted sequence of rings with adjacent rings having one atom in common (**spiro polymer**) or two or more atoms in common (**ladder polymer**).

The source-based nomenclature is based upon the starting monomer(s) from which the ladder or spiro polymer was prepared.

Example

1.

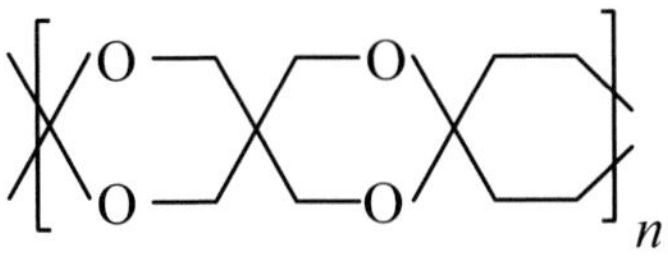

Source-based name: *spiro*-poly[2,2-bis(hydroxymethyl)propane-1,3-diol-*alt*▶
-cyclohexane-1,4-dione]

Example 2 demonstrates the major difference between source-based and structure-based nomenclature styles.

Example

2.

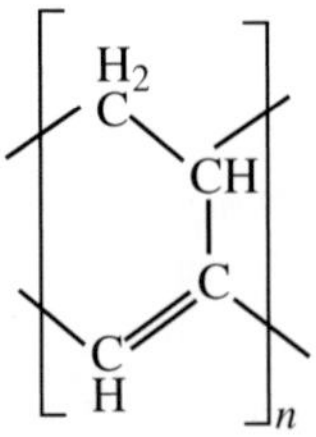

The structure-based name is poly(but-1-ene-1,4:3,2-tetrayl), but the source-based name, *ladder*-poly(methyl vinyl ketone), can only be recognised with knowledge of the reactions involved in its formation. In the generation of this polymer, the monomer, methyl vinyl ketone, has to undergo chain polymerisation, followed by a cyclic elimination of water between keto and methyl groups. The polymer bears little structural resemblance to the monomer from which it is formed.

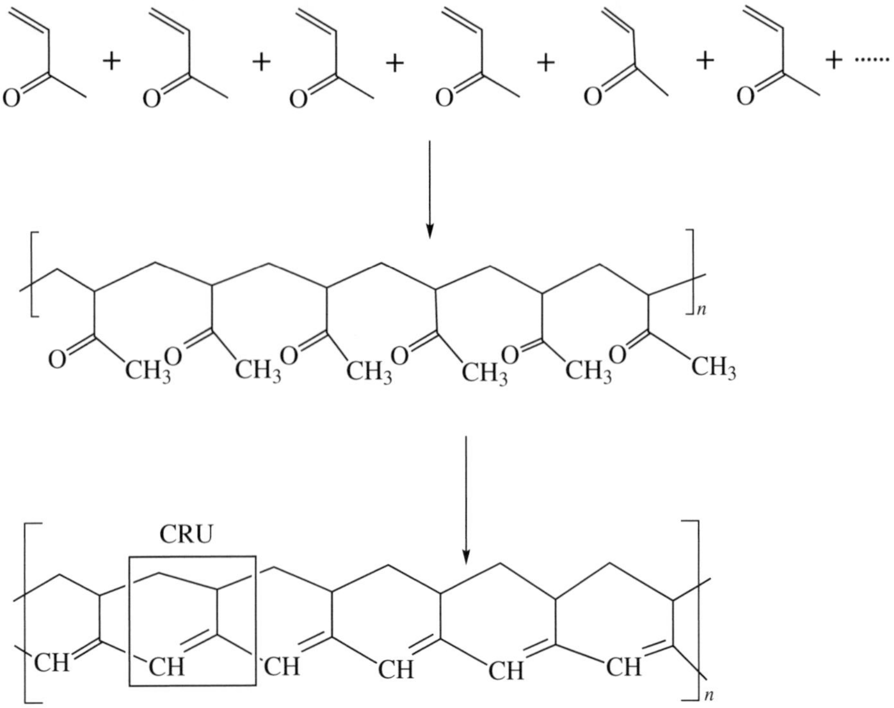

Example 2 illustrates the problems that can occur when a source-based name for a polymer (product) is used, the formation of which has involved several isolable intermediates in the course of the reaction.

A structure-based name can also be derived for double-strand polymers, using all the rules described above for the determination of the CRU and seniority for single-strand polymers. However, the CRU is usually a tetravalent group denoting attachment to four atoms and it is named according to the usual rules of the nomenclature of organic chemistry. The name of the polymer is then written in the form poly(CRU). Because a double-strand polymer has a sequence of rings, in order to identify a single preferred CRU, the rings must be broken formally by observing the following criteria in decreasing order of priority.

1. The number of free valences in the CRU is minimised.
2. The number of most-preferred heteroatoms in the ring system is maximised.
3. The most preferred ring system is retained.
4. For an acyclic CRU, the longest chain is chosen.

Further decisions are based on the seniority of ring systems (for details see Chapter 6), on the orientation of the CRU so as to position the lowest free-valence locant at the lower left of the structural diagram as written, and on placing the acyclic subunits, if any, within the CRU to the right of the ring system. In the name the locants for the free valences are placed directly in front of the corresponding suffix and cited in clockwise order beginning from the lower left position in the formula, the locants for the left positions being separated from the right ones by a colon.

Examples

3. A polymer consisting of adjacent six-membered saturated carbon rings:

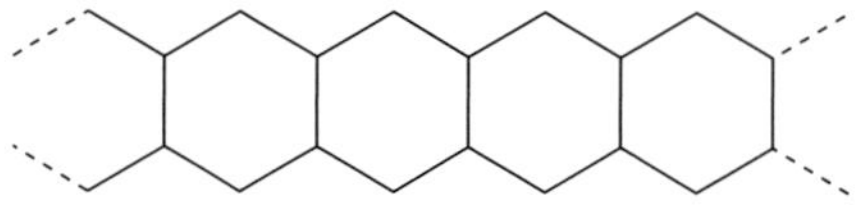

For this polymer, the preferred CRU is assigned as shown.

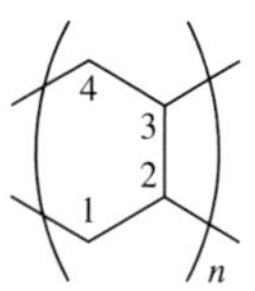

The name is therefore poly(butane-1,4:3,2-tetrayl).

4. A polymer consisting of adjacent cyclohexane and 1,3-dioxane rings in a regular spiro sequence:

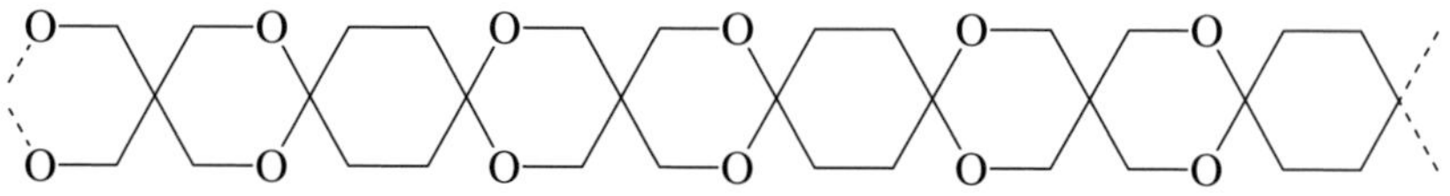

The structure-based name is derived from the preferred CRU shown below.

The structure-based name is therefore poly[2,4,8,10-tetraoxaspiro[5.5]►undecane-3,3:9,9-tetrayl-9,9-di(ethane-1,2-diyl)]. For the source-based name, see Example 1 of this Section.

9.6 NOMENCLATURE OF REGULAR SINGLE-STRAND INORGANIC AND COORDINATION POLYMERS (reference 12)

The names of regular single-strand inorganic and coordination polymers are based on the same fundamental principles and assumptions that were developed for regular single-strand organic polymers. As in the nomenclature of organic polymers the name of the polymer is the name of the CRU prefixed by 'poly'. However, inorganic substances may often be named using either additive nomenclature or substitutive nomenclature, whereas organic polymer names are generally derived using substitutive nomenclature, as described above. When using the additive approach for naming inorganic polymers the CRU is in general enclosed in square brackets. In addition, the name is preceded by an italic structural descriptor such as '*catena*'. This does not imply a catenane but the presence of a chain structure.

Examples

1.

catena-poly[dimethyltin]
(additive style name, see Chapter 4) **or**
poly(dimethylstannanediyl)
(substitutive style name, see Chapter 4)

2.

catena-poly[(difluoridosilicon)(dimethylsilicon)]
(additive style name, see Chaper 4) **or**
poly(1,1-difluoro-2,2-dimethyldisilane-1,2-diyl)
(substitutive style name, see Chapter 4)

The names 'silicone' (not to be confused with the name silicon for the element) and 'polysiloxane' are sometimes used for polymers composed of macromolecules containing alternating oxygen and silicon atoms in the backbone (Example 3).

Example

3.

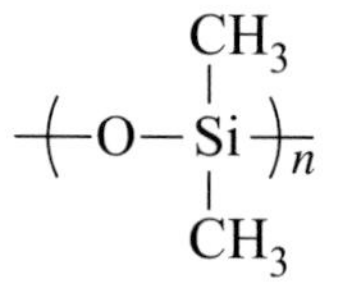

$$\left(\!\!-O-\underset{\underset{CH_3}{|}}{\overset{\overset{CH_3}{|}}{Si}}\!\!\right)_{\!n}$$

catena-poly[(dimethylsilicon)-μ-oxido]
(additive style name) **or**
poly[oxy(dimethylsilanediyl)]
(substitutive style name) **or**
poly(dimethylsiloxane)
(traditional name, still used on occasion) **or**
dimethylsilicone
(commonly used)

Examples 1 and 2 above are regular, single-strand, inorganic polymers with simple homoatomic backbones. However, coordination polymers commonly consist of single central metal atoms with bridging ligands. In order to select the preferred CRU, the seniorities of its subunits are considered, as well as the preferred direction for the sequential citation. The constituent subunit of highest seniority must contain one or more central atoms. Bridging ligands between central atoms in the backbone of the polymer cannot be senior subunits. This is consistent with the principle of coordination nomenclature, in which the emphasis is laid on the coordination centre. Such a polymer is therefore named by citing the central atom, prefixed by its associated non-bridging ligands, followed by the name of the bridging ligand, which in turn is prefixed by the Greek letter μ (see Example 4 and also the additive name in Example 3).

Example

4.

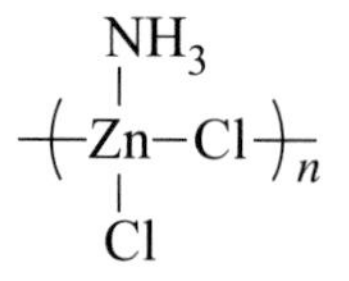

$$\left(\!\!-\underset{\underset{Cl}{|}}{\overset{\overset{NH_3}{|}}{Zn}}\!\!-Cl\!\!\right)_{\!n}$$

catena-poly[(amminechloridozinc)-μ-chlorido]

In inorganic additive nomenclature, a regular polymer that can be described by a preferred CRU in which only one terminal constituent subunit is connected through a single atom (the silicon atom in Example 5 below) to the other identical CRU is viewed as a quasi-single-strand polymer. Such polymers are named in a fashion similar to that used for single-strand coordination polymers.

Example

5.

$$\left[Si \begin{array}{c} S \\ S \end{array} \right]_n$$

catena-poly[silicon-di-μ-sulfido]
(additive style name) **or**
poly[silanetetraylbis(sulfanediyl)]
(substitutive style name)

9.7 ABBREVIATIONS FOR POLYMER NAMES

The International Organization for Standardization (ISO) published a revised list of abbreviations for polymer names in 2001 (reference 13) and the list of abbreviations in Table 9.2 is derived from that list, though it also contains earlier IUPAC recommendations. However, the ISO list uses nomenclature that is not necessarily in accord with IUPAC recommendations (reference 14).

The IUPAC recommendation on the use of abbreviations in the chemical literature (reference 15) declares that: 'there are great advantages in defining all abbreviations ... in a single conspicuous place in each paper. This is preferably done near the beginning of the paper in a single list.' Each abbreviation should therefore be fully defined the first time it appears in the text, and no abbreviation should be used in titles or abstracts of publications.

9.8 NAMES OF COMMON CONSTITUTIONAL UNITS IN POLYMERS

Constitutional units are used in construction of structure-based polymer names, as discussed in Section 9.2.1. Table P9 lists many of the constitutional units commonly encountered in polymer names. The names in the left-hand column and marked with an asterisk (*) should not be used, as they are no longer recommended (references 7 and 16). These are included because they are still found in the literature. For such cases the correct name is in the right-hand column. Those entries also marked † are retained IUPAC names for groups other than the ones being indicated here or for the same group but in a different context. Unfortunately, this may give rise to some confusion. Table P9 is based upon the material in reference 1.

9.9 TRADITIONAL NAMES FOR COMMON POLYMERS

A number of common polymers have semi-systematic or trivial names that are well established by usage. For scientific communication, the use of source-based polymer names containing trivial monomer names should be kept to a minimum. For the idealised structural representations included in Table 9.1, semi-systematic or trivial source-based names are given and the corresponding structure-based names are alternatives. An overview of polymer classes and their names including those of inorganic polymers (for example, phosphorus-containing polymers) is given in reference 17.

Table 9.2 Abbreviations of polymer names

ABS	poly(acrylonitrile-*co*-buta-1,3-diene-*co*-styrene)
CA	cellulose acetate
CMC	(carboxymethyl)cellulose
EVAC	poly[ethene-*co*-(vinyl acetate)]
PA	polyamide
PAN	polyacrylonitrile
PB	polybutene (general, there are several forms)
PBAK	poly(butyl acrylate)
PBT	poly(butane-1,4-diyl terephthalate)
PC	polycarbonate
PE	polyethene; polyethylene
PEO	poly(ethylene oxide)
PET	poly(ethylene terephthalate)
PMMA	poly(methyl methacrylate)
POM	poly(oxymethylene); polyformaldehyde
PP	polypropene
PS	polystyrene
PTFE	poly(tetrafluoroethene)
PUR	polyurethane[a]; polycarbamate
PVAC	poly(vinyl acetate)
PVAL	poly(vinyl alcohol)
PVC	poly(vinyl chloride)
PVDF	poly(1,1-difluoroethene); poly(vinylidene difluoride)[a]
PVK	poly(9-vinylcarbazole); poly(*N*-vinylcarbazole)
PVP	poly(1-vinylpyrrolidone); poly(*N*-vinylpyrrolidone)
UP	unsaturated polyester

[a]Name upon which the abbreviation is based.

REFERENCES

1. *Compendium of Polymer Terminology and Nomenclature*, RSC Publishing, Cambridge, UK, 2009. This is the second edition of the so-called Purple Book.
2. Glossary of basic terms in polymer science, *Pure and Applied Chemistry* 1996, **68**(12), 2287–2311. Reprinted as Chapter 1 in reference 1.
3. Reference 1, Chapter 14.
4. Source-based nomenclature for copolymers, *Pure and Applied Chemistry*, 1985, **57**(10), 1427–1440. Reprinted as Chapter 19 in reference 1.
5. Source-based nomenclature for non-linear macromolecules and macromolecular assemblies, *Pure and Applied Chemistry*, 1997, **69**(12), 2511–2521. Reprinted as Chapter 20 in reference 1.
6. Structure-based nomenclature for cyclic organic macromolecules, *Pure and Applied Chemistry*, 2008, **80**(2), 201–232.
7. Nomenclature of regular single-strand organic polymers, *Pure and Applied Chemistry*, 2002, **74**(10), 1921–1956. Reprinted as Chapter 15 in reference 1.
8. Structure-based nomenclature for irregular single-strand organic polymers, *Pure and Applied Chemistry*, 1996, **66**(4), 873–879. Reprinted as Chapter 17 in reference 1.
9. Nomenclature of fused and bridged fused ring systems, *Pure and Applied Chemistry*, 1998, **70**(1), 143–216.
10. Generic source-based nomenclature for polymers, *Pure and Applied Chemistry*, 2001, **73**(9), 1511–1519. See also errata in *Pure and Applied Chemistry*, 2002, **74**(10), 2019. Reprinted as Chapter 21 in reference 1.
11. Nomenclature of regular double-strand (ladder and spiro) organic polymers, *Pure and Applied Chemistry*, 1993, **65**(7), 1561–1580. Reprinted as Chapter 16 in reference 1.
12. Nomenclature for regular single-strand and quasi single-strand inorganic and coordination polymers. *Pure and Applied Chemistry*, 1985, **57**(1), 149–168.
13. ISO International Standard, ISO 1043-1:2001. *Plastics-Symbols and Abbreviated Terms* – Part 1: Basic polymers and their special characteristics.
14. Reference 1, Chapter 22.
15. Use of abbreviations in the chemical literature (Recommendations 1979), *Pure and Applied Chemistry*, 1980, **52**(9), 2229–2232.
16. *A Guide to IUPAC Nomenclature of Organic Compounds (Recommendations 1993)* Blackwell Scientific Publications, Oxford, See also Errata, *Pure and Applied Chemistry*, 1999, **71**(7), 1327–1330.
17. Glossary of class names of polymers based on chemical structure and molecular architecture, *Pure and Applied Chemistry*, 2009, **81**(6), 1131–1186.

10 Boron Hydrides and Related Compounds

10.1 BORON HYDRIDES AS A UNIQUE GROUP OF COMPOUNDS

10.1.1 Introduction

Boron hydrides and related compounds possess unusual structural features and, as a direct consequence, they present their own unique nomenclature challenges. A specific nomenclature has developed to describe them, and this has aspects that differ from the more generalised nomenclatures that have been discussed so far. An example is the use of the term 'hydro' where IUPAC inorganic nomenclature generally recommends the use of 'hydrido'.

Boron hydrides are often described as 'electron deficient', but this is clearly a misnomer, since many are not the strong Lewis acids they would be expected to be if they were really electron deficient. Boron atoms in boron hydrides avoid this so-called 'electron deficiency' by partaking freely in multicentre bonding, with the consequence that triangulated polyhedral structures or clusters are not unusual. Although the system which has been developed for naming boron hydrides and their derivatives appears at first sight to be very specialised, it is essentially an adapted extension of the current IUPAC nomenclature systems. This chapter is an introduction to this nomenclature. More detailed, comprehensive treatments may be found elsewhere (references 1–4).

10.1.2 Names of simple boranes and boron hydride anions

Compositional (stoichiometric) names of binary boron hydride species are easy to derive, and follow the principles described earlier (Chapter 5). Neutral binary boron hydrides are named **boranes**, with a prefix to indicate the number of boron atoms present, followed by the number of hydrogen atoms in parentheses. Examples 1 and 2 illustrate these names.

Examples

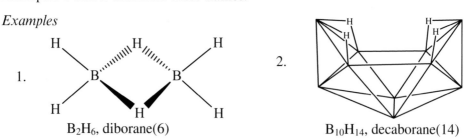

1. B_2H_6, diborane(6)

2. $B_{10}H_{14}$, decaborane(14)

Example 2 is a **polyhedral polyborane**. Rather like the practice in organic usage where carbon atoms are not shown explicitly in a structure, in this case the

Principles of Chemical Nomenclature: A Guide to IUPAC Recommendations
Edited by G. J. Leigh
© International Union of Pure and Applied Chemistry 2011
Published by the Royal Society of Chemistry, www.rsc.org

boron atoms are not shown explicitly, but there is a boron atom at every vertex of the polyhedron, each bound to an *exo* hydrogen atom, likewise not shown explicitly. However, there are additional bridging H atoms in this structure at the positions shown. The structures of the polyhedral boranes are treated in more detail in Section 10.2.

Anions derived from these boranes are named by replacing the 'ane' ending with the anion suffix 'ate', as discussed more generally in Chapter 5. The charge is indicated in parentheses after the name, also as described more generally elsewhere (Section 5.4), with the number of boron atoms specified as in neutral boranes. However, it is always necessary to indicate the number of hydrogen atoms, but this is not done as with the neutral boranes, but by placing a numerical prefix (Table P6) indicating the number of hydrogen atoms at the beginning of the name, followed without a break or hyphen by a second prefix, 'hydro'. However, the *Nomenclature of Inorganic Chemistry, IUPAC Recommendations 2005*, advises that 'hydrido' would be preferable in such circumstances. Examples 3 and 4 illustrate names derived in this manner.

Examples

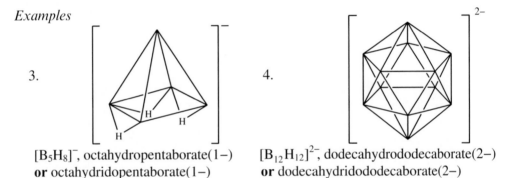

3.

4.

$[B_5H_8]^-$, octahydropentaborate(1−)
or octahydridopentaborate(1−)

$[B_{12}H_{12}]^{2-}$, dodecahydrododecaborate(2−)
or dodecahydridododecaborate(2−)

10.2   STRUCTURAL CONSIDERATIONS AND POLYHEDRA

The structural basis of boron hydride chemistry needs to be considered before a more detailed systematic nomenclature can be developed. Boron hydrides often adopt structures related to regular triangulated polyhedra (**deltahedra**) and these polyhedral structures are termed clusters. The common triangulated polyhedral shapes are shown in Figure 10.1. In the simpler binary boranes, boron atoms occupy the vertices of the polyhedral shape. The lines in Figure 10.1 are drawn to represent the polyhedral shapes and do not imply the locations of electron-pair bonds. Electron-counting rules have been developed (reference 5) to enable the structures of boron hydrides (and their related compounds) to be understood and rationalised from their molecular (or ionic) formulae. The structures of boron hydrides are directly related either to polyhedra as drawn in Figure 10.1 (called *closo* structures) or to more open structures which are best represented as being derived from the *closo* polyhedra with either one vertex missing (giving a *nido* structure) or two vertices missing (giving an *arachno* structure). It is usually the vertex of highest connectivity (bonded to the greatest number of neighbours) which is removed to convert a *closo* structure to a *nido* structure, and an additional vertex adjacent to the highest connected vertex is also removed to give rise to an *arachno* structure.

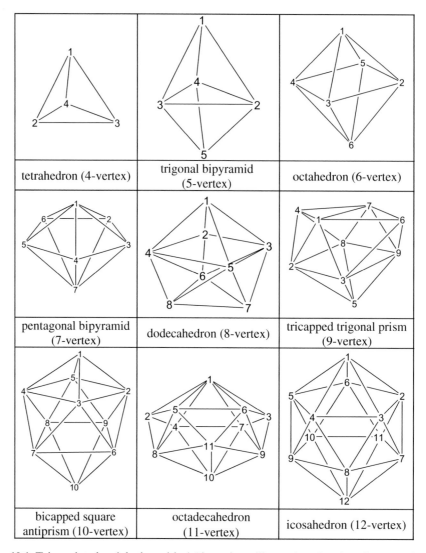

Figure 10.1 Triangulated polyhedra with 4-12 vertices, illustrating the *closo* framework cluster geometries, and showing numbering conventions.

In the simpler binary boranes, H atoms are attached to each boron atom by bonds which are directed radially out from the centre of the polyhedron, in terminal or *exo* positions. In all the diagrams in this Chapter, unmarked vertices represent such {BH} units. In the more open *nido/arachno* structures, additional H atoms are also present and these, called *bridging* hydrogen atoms, either bridge two boron vertex atoms on an open face, or they may be associated solely with one boron atom and oriented towards a missing vertex, when they are described as *endo* terminal H atoms. The geometric shapes of the more commonly observed open structures are shown in Figure 10.2. These structural descriptors *nido* and *arachno* are of prime significance in boron hydride chemistry and when specified they enable the geometry/structure of a given borane cluster to be envisaged. For example, a binary boron hydride, with *n* boron atoms which is described as *nido*, is based on the regular triangulated *closo* $(n + 1)$ vertex polyhedral shape, but with the highest connectivity vertex removed. One which is described as *arachno* is based on the regular triangulated

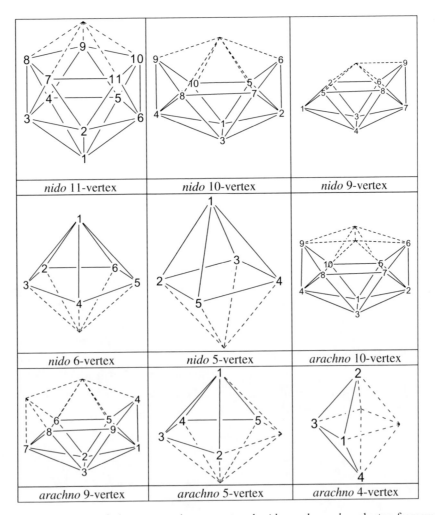

Figure 10.2 Structures of the commonly encountered *nido-* and *arachno*-cluster frameworks, showing their numbering conventions.

closo $(n + 2)$ vertex polyhedral shape with two adjacent vertices removed. These structural descriptors are usually inserted into the name immediately before the descriptor indicating the number of boron atoms (see Examples 1 and 2).

Examples

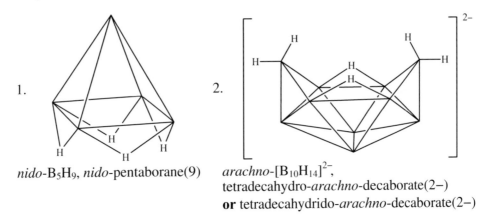

1. *nido*-B_5H_9, *nido*-pentaborane(9)

2. *arachno*-$[B_{10}H_{14}]^{2-}$,
tetradecahydro-*arachno*-decaborate(2−)
or tetradecahydrido-*arachno*-decaborate(2−)

Other structural descriptors which are less commonly used include *hypho*, *conjuncto*, and *commo*. They indicate more complex structures and though all are used in practice only *hypho* has yet received IUPAC approval.

10.3 SUBSTITUTION AND REPLACEMENT IN BORON CLUSTERS

10.3.1 H atom replacement

Boron hydride clusters are typified as having polyhedra of boron atoms with triangulated faces, each boron atom carrying an *exo* hydrogen atom. Substitution of hydrogen atoms by various functional groups is possible, to produce substituted compounds such as organoboranes and halogenoboranes. To avoid ambiguity and to define these derivatives accurately, numbering schemes for boron vertices (and associated *exo* substituents) have been developed. Bridging hydrogen atoms may also be substituted and these are indicated using the bridging symbol μ and numbering the two boron vertices that are bridged (compare Section 7.2.5). The numbering schemes for commonly observed cluster geometries with between 4 and 12 vertices are shown in Figures 10.1 and 10.2. The replacing group may be either anionic (Example 1) or neutral (Example 2). Neutral groups replacing *exo*-H atoms are considered as ligands to the polyhedral borane cluster, and these generally increase the number of electrons available for skeletal bonding. This causes a rearrangement of the cluster to give a more open structure.

Examples

1.

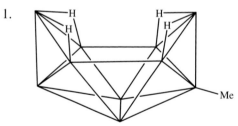

2-Me-*nido*-$B_{10}H_{13}$, 2-methyl-*nido*-decaborane(14)

2.

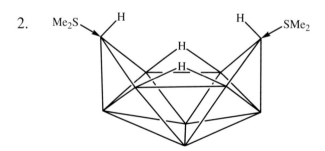

6,9-$(Me_2S)_2$-*arachno*-$B_{10}H_{12}$, 6,9-bis(dimethylsulfane)-*arachno*-decaborane(12)

The arrows in Example 2 are used here simply to aid electron counting, and are not recommended for general use.

10.3.2   Subrogation (skeletal replacement)

Boron atoms within a polyhedral cluster framework (skeleton) may be replaced by other Main Group elements or by transition elements with the structural integrity of the cluster being retained. Such skeletal replacement is called subrogation. The term subrogation is specific to skeletal replacements in boron hydride chemistry, and is not used elsewhere. The atom which subrogates the boron atom often (but not necessarily) has *exo* groups associated with it; *exo* groups may be hydrogen atoms, halogens, organic groups, ligands, *etc.* Subrogated atoms are indicated using the conventional replacement-names system (see Section 6.3.3 and Table 6.2), *e.g.*, carba represents C, aza N, oxa O, sila Si, phospha P, thia S, arsa As, but the name of the substituted compound is derived from the parent unsubrogated borane. Boranes with boron atoms subrogated by Main Group atoms are classified as **heteraboranes** whilst those which have boron atoms subrogated by transition-metal atoms are termed **metallaboranes**. The terms heteroborane(s) and metalloborane(s) are also in common usage, though heteraborane(s) and metallaborane(s) are preferred in practice. The names for specific metal atoms incorporated into the cluster follow the conventional system with the proviso that when there is a choice of numbering then, as usual, the heteroatom is assigned the lowest possible locant.

Metallaboranes, heteraboranes, and metallaheteraboranes, are not uncommon. Adopting the procedures described above will usually result in unambiguous names for compounds. In Example 3 the sulfur atom has an *exo* lone pair, which need not necessarily be shown as it is here. In Example 4 there are *exo* hydrogen atoms (unspecified) on each carbon vertex.

Examples

3.
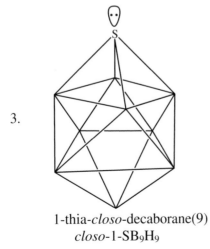

1-thia-*closo*-decaborane(9)
closo-1-SB$_9$H$_9$

4.
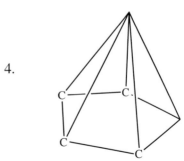

2,3,4,5-tetracarba-*nido*-hexaborane(6)
nido-2,3,4,5-C$_4$B$_2$H$_6$

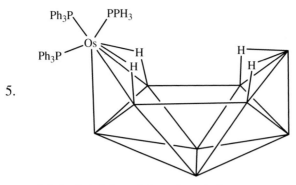

5.

6,6,6-tris(triphenylphosphane)-6-osma-*nido*-decaborane(13)
6,6,6-(PPh$_3$)$_3$-*nido*-6-OsB$_9$H$_{13}$

Trivial names are also common in metallaborane, heteroborane, and metallaheteraborane chemistry, and important examples are included below. Boron hydrides with one or more boron atoms subrogated by carbon atoms (and their derived anions) are an important class of compound which is often referred to generically as carboranes (or carborane anions). According to IUPAC recommendations they should be known as carbaboranes (or carbaborane anions). Anions are named in accordance with the procedure shown in Section 10.1.2. In Examples 6–10 each carbon atom has an (unmarked) *exo* hydrogen atom associated with it.

Examples

6.
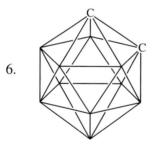

closo-1,2-C$_2$B$_{10}$H$_{12}$
1,2-dicarba-*closo*-dodecaborane(12)

7.
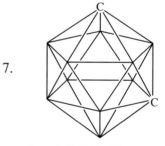

closo-1,7-C$_2$B$_{10}$H$_{12}$
1,7-dicarba-*closo*-dodecaborane(12)

8.
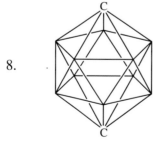

closo-1,12-C$_2$B$_{10}$H$_{12}$
1,12-dicarba-*closo*-dodecaborane(12)

9.
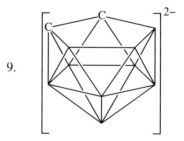

[*nido*-7,8-C$_2$B$_9$H$_{11}$]$^{2-}$
undecahydro-7,8-dicarba-*nido*-▶
undecaborate(2–)

or undecahydrido-7,8-dicarba-*nido*-▶
undecaborate(2–)

1,2-Dicarba-*closo*-dodecaborane(12),   1,7-dicarba-*closo*-dodecaborane(12), and 1,12-dicarba-*closo*-dodecaborane(12) are often referred to by their trivial names of *ortho*-carborane, *meta*-carborane, and *para*-carborane, though these names are not approved by IUPAC. The dianion, $[nido\text{-}7,8\text{-}C_2B_9H_{11}]^{2-}$, has a rich co-ordination chemistry and is sometimes referred to as the **carbollide(2–)** ligand. The carbollide ligand bears more than a superficial resemblance to the cyclopentadienide(1–) anion. However, since the metal centre is effectively now part of the cluster, such compounds should be named as subrogated boron hydride clusters; in older literature such compounds are often described as η^5-complexes of the $[nido\text{-}7,8\text{-}C_2B_9H_{11}]^{2-}$ anion (Example 10).

Examples

10.

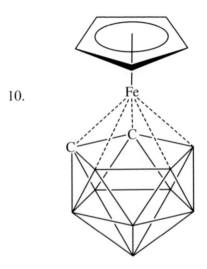

3-(η^5-C_5H_5)-*closo*-1,2,3-$C_2FeB_9H_{11}$
3-(η^5-cyclopentadienyl)-1,2-dicarba-3-ferra-*closo*-dodecaborane(11)

Note the use (in this text alone) of the symbol ▶ in Example 9 to indicate a break imposed by typographical considerations, though this symbol is not part of the name string.

REFERENCES

1. R. M. Adams, Nomenclature of inorganic boron compounds, *Pure and Applied Chemistry*, 1972, **30**, 683–710.
2. *Nomenclature of Inorganic Chemistry, Recommendations 1990*, Chapter I-11, Blackwell Scientific Publications, Oxford, 207–237.
3. *Principles of Chemical Nomenclature, A Guide to IUPAC Recommendations*, G. J. Leigh, H. A. Favre, and W. V. Metanomski, Blackwell Science, Oxford, 1998.
4. *Nomenclature of Inorganic Chemistry, IUPAC Recommendations 2005*, Chapter IR-6, Royal Society of Chemistry, Cambridge, 83–110.
5. K. Wade, Structural and bonding patterns in cluster chemistry, *Advances in Inorganic Chemistry and Radiochemistry*, 1976, **18**, 1–66.

11   Biochemical Nomenclature

11.1   INTRODUCTION

11.1   INTRODUCTION

Systematic substitutive nomenclature (Chapter 6) may be used to name all organic molecules. However, those that are of animal or vegetable origin are often complex and have received trivial names, such as cholesterol, oxytocin and glucose. Much, but not all, of the nomenclature described in Chapter 6 may be applied to these materials. Biochemical nomenclature is based upon these trivial names, which are either substitutively modified in accordance with the principles, rules and conventions described in Chapter 6, or transformed and simplified into names of parent structures, *i.e.*, parent compounds of a specific configuration. These names are then modified by the rules of substitutive nomenclature. Four classes of compound will be discussed here to illustrate the basic approach: carbohydrates; amino acids and peptides; lipids; and nucleic acids. For further detail, see *Biochemical Nomenclature and Related Documents*, 2nd Edition, Portland Press, London, 1992 or individual documents on the Web.

11.2   CARBOHYDRATE NOMENCLATURE

Originally, **carbohydrates** were defined as compounds, such as aldoses and ketoses, having the empirical, stoichiometric formula $C(H_2O)$, hence 'hydrates of carbon'. The generic term '**carbohydrates**' includes **monosaccharides**, **oligosaccharides** and **polysaccharides**, as well as substances derived from monosaccharides by reduction of the carbonyl group (**alditols**), by oxidation of one or more terminal groups to carboxylic acid(s), or by replacement of one or more hydroxy group(s) by a hydrogen atom, an amino group, or a sulfanyl group or similar group. It also includes derivatives of these compounds. The term carbohydrate is synonymous with the term **saccharide**. Trivial names are common in carbohydrate nomenclature and fifteen of them form the basis of the systematic nomenclature. They are assigned to the simple aldoses (polyhydroxyaldehydes), from triose to hexoses. These class names are based upon the number of carbon atoms each contains, from three to six.

Triose: glyceraldehyde (*not* glycerose)
Tetroses: erythrose, threose
Pentoses: arabinose, lyxose, ribose, xylose
Hexoses: allose, altrose, galactose, glucose, gulose, idose, mannose, talose

In addition to the aldoses, the hexoses also include compounds which contain a ketone rather than an aldehyde group. These are called **ketohexoses**, and among the ketohexoses **fructose** is of major natural occurrence.

Principles of Chemical Nomenclature: A Guide to IUPAC Recommendations
Edited by G. J. Leigh
© International Union of Pure and Applied Chemistry 2011
Published by the Royal Society of Chemistry, www.rsc.org

A stereodescriptor, D or L, is used to indicate the absolute configuration of the entire molecule. Reference is made to the configuration of glyceraldehyde through a reference carbon atom, which is the carbon atom of the stereogenic centre receiving the highest numerical locant (marked by arrows in the Examples 1–5), the lowest possible number being given to the carbon atom bearing the principal characteristic group, which is a carbonyl group in aldoses and ketoses.

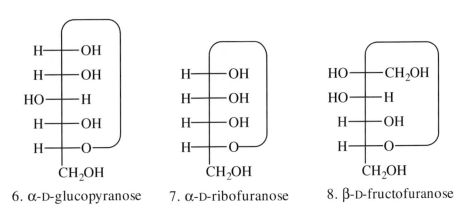

D-glyceraldehyde

Examples

1. D-glyceraldehyde 2. D-threose 3. D-ribose 4. D-glucose 5. D-fructose

The ketone and aldehyde groups can eliminate water by reacting with an alcohol group in the same molecule, yielding cyclic molecules. The names of such cyclised (**hemi-acetalised**) aldoses and ketoses contain the infixes pyran or furan to indicate the six- or five-membered heterocyclic structure, and a stereodescriptor, α or β, to indicate the configuration of the **anomeric** or hemi-acetal carbon atom. Examples 6–8 show three such cases

Examples

6. α-D-glucopyranose 7. α-D-ribofuranose 8. β-D-fructofuranose

Although the names of the saccharides are generally trivial, systematic nomenclature is used to name their derivatives. Because trivial names are not amenable to the treatments usually applied to the names of ordinary parent hydrides, many adaptations are necessary and some peculiarities must be noted, for example, substitution can be made on an oxygen atom in the case of esters and ethers. This substitution is characterised by the symbol *O*, which is placed after the locant. The compound prefix 'deoxy' is composed of the prefixes 'de', meaning 'without' in subtractive nomenclature, and 'oxy', to indicate the subtraction of an oxy group from an –OH group: C–O–H→C–H. Such an operation is needed to indicate the replacement of an –OH group by another group, such as an amino group.

Examples

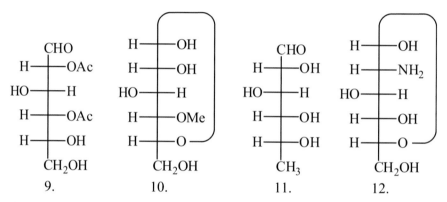

9. 2,4-di-*O*-acetyl-D-glucose [acetyl = –C(O)–CH₃]

10. 4-*O*-methyl-α-D-glucopyranose

11. 6-deoxy-D-glucose

12. 2-amino-2-deoxy-α-D-glucopyranose

Acid and alcohol derivatives are named by changing the ending 'ose' of the saccharide name into the appropriate ending to signify a functional change, for instance, 'onic acid' or 'aric acid' or 'uronic acid', and 'itol'.

Examples

13.	X = CHO	Y = CH₂OH	D-glucose
14.	X = COOH	Y = CH₂OH	D-gluconic acid
15.	X = COOH	Y = COOH	D-glucaric acid
16.	X = CHO	Y = COOH	D-glucuronic acid
17.	X = CH₂OH	Y = CH₂OH	D-glucitol

The generic term **glycoside** defines all mixed acetals formed by the acetalisation of the cyclic forms of aldoses and ketoses. **Glycosyl groups** are the products obtained when monosaccharides lose their anomeric –OH group; the suffix 'yl' is used to indicate the change that has occurred at C-1.

Examples

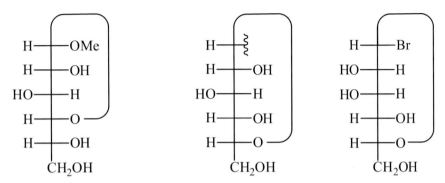

18. methyl α-D-glucofuranoside 19. α-D-glucopyranosyl   20. α-D-mannopyranosyl bromide

Disaccharides are named by adding the name of a glycosyl group as a prefix to that of the monosaccharide chosen as parent, as exemplified by β-lactose. Example 21 shows both a Haworth perspective formula (a) and a conformational formula (b).

Example

21.

(a)   (b)

β-D-galactopyranosyl-(1→4)-α-D-glucopyranose

Two forms of abbreviated nomenclature (extended and short) may be used; D-glucopyranose is represented as D-Glc*p* in the extended form and Glc in the condensed form. The linking atoms are designated by locants, and the α or β configuration of the anomeric carbon atoms is also indicated. The sugar raffinose becomes α-D-Gal*p*-(1→6)-α-D-Glc*p*-(1↔2)-β-D-Fru*f* (extended form) or Galα-6Glcα-βFru*f* (short form).

Aldoses are systematically named as pentoses, hexoses, heptoses, octoses, nonoses, *etc.*, according to the total number of carbon atoms in the chain. The configuration is described by appropriate stereodescriptors (*glycero-* from glyceraldehyde, *gluco-* from glucose, *galacto-* from galactose, *etc.*) together with the appropriate D or L, and these are assembled in front of the basic name according to specific rules. Names of ketoses are characterised by ending in 'ulose'.

Examples

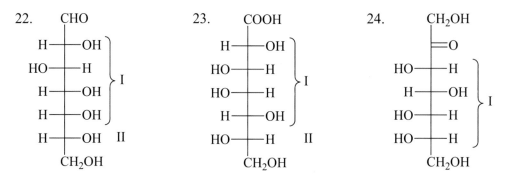

22. D-*glycero*-D-*gluco*-heptose (I = D-*gluco*, II = D-*glycero*)
23. L-*glycero*-D-*galacto*-heptaric acid (I = D-*galacto*, II = L-*glycero*)
24. L-*gluco*-hept-2-ulose (I = L-*gluco*)

11.3 NOMENCLATURE AND SYMBOLISM FOR α-AMINO ACIDS AND PEPTIDES

Although there are many **amino acids** in Nature, the term amino acid is usually restricted to mean α-amino acids (2-aminocarboxylic acids), and, in particular, the 20 α-amino acids incorporated into proteins under mRNA direction (Table 11.1).

A **peptide** is any compound produced by amide formation between a carboxy group of one amino acid and an amino group of another. The amide bonds in peptides are called peptide bonds. The word peptide is usually applied to compounds whose amide bonds (sometimes called **eupeptide** bonds) are formed between C-1 of one amino acid and N-2 of another, but it includes compounds with residues linked by other amide bonds (sometimes called **isopeptide** bonds). Peptides with fewer than about 10–20 residues may also be called **oligopeptides**; those with more residues are called **polypeptides**. Polypeptides of specific sequence of more than about 50 residues are usually known as **proteins**, but authors differ greatly on where they start to apply this term.

Amino acids are represented in two ways: either as H_2N–CHR–COOH or as the zwitterionic form H_3N^+–CHR–COO$^-$. Although the second of these forms is

Table 11.1 Names and their abbreviations and symbols for common amino acids found in proteins*

Trivial name	Symbols		Trivial name	Symbols	
Alanine	Ala	A	Leucine	Leu	L
Arginine	Arg	R	Lysine	Lys	K
Asparagine	Asn	N	Methionine	Met	M
Aspartic acid	Asp	D	Phenylalanine	Phe	F
Cysteine	Cys	C	Proline	Pro	P
Glutamine	Gln	Q	Serine	Ser	S
Glutamic acid	Glu	E	Threonine	Thr	T
Glycine	Gly	G	Tryptophan	Trp	W
Histidine	His	H	Tyrosine	Tyr	Y
Isoleucine	Ile	I	Valine	Val	V

The general representation for an unspecified amino acid is Xaa, symbol X.

overwhelmingly predominant in the crystal and in solution, it is generally more convenient to name amino acids and their derivatives from the first form. They are normal organic compounds and are treated as such as far as numbering and naming are concerned, although trivial names are retained for all natural amino acids.

Examples

1. $H_2N–CH_2–COOCH_3$ methyl glycinate **or** glycine methyl ester

2. $CH_3–CH(NH_2)–CONH_2$ alaninamide **or** alanine amide
3. $HOCH_2–CH(NHCOCH_3)–COOH$ *N*-acetylserine

There are two generally accepted systems of abbreviation for trivial names, using either one or three letters. The choice of which form to use is generally determined by circumstances. Normally three-letter symbols are used, and the one-letter symbols are reserved for long sequences of L-amino acids. A list of such symbols is shown in Table 11.1. The stereodescriptors D and L are used with reference to the D configuration of glyceraldehyde.

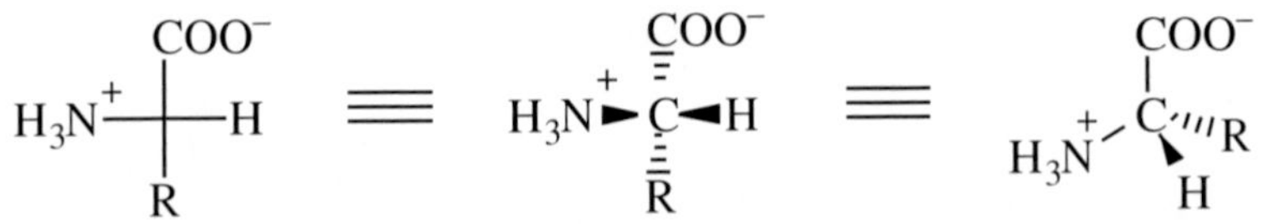

Three-letter symbols and standard group abbreviations are used to designate amino acids functionalised on –COOH or substituted on –NH₂. In abbreviations modifications are indicated by hyphens inserted between the symbols of the substituent and the parent. The symbol is assumed to represent the amino group on the left and carboxy group on the right.

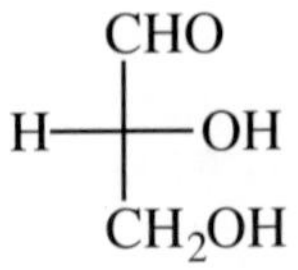

Examples

4. *N*-acetylglycine   Ac-Gly
5. ethyl glycinate   Gly-OEt
6. N^2-acetyllysine   Ac-Lys
7. O^1-ethyl *N*-acetylglutamate   Ac-Glu-OEt

Substitution on any other part of the amino acid is expressed by a vertical line to the substituent.

Example

8. *S*-ethylcysteine
$$\overset{\text{Et}}{\overset{|}{\text{Cys}}} \quad \textbf{or} \quad \text{Cys(Et)}$$

The peptide $H_2N-CH_2-CO-NH-CH(CH_3)-COOH$ is named *N*-glycylalanine and symbolised as Gly-Ala. The amino acid with the free –COOH group is chosen as the parent. The name of the other amino acid, modified by the suffix 'yl', becomes a prefix to it.

The symbolism applied to a peptide is very precise and elaborate. The symbol -Ala stands for $-NH-CH(CH_3)-COOH$, and the corresponding name is that of the amino acid. The symbol Gly- means H_2N-CH_2-CO- and corresponds to a name ending in 'yl'. In the peptide Gly-Gly-Ala, Gly- signifies H_2N-CH_2-CO- and -Gly- signifies $-HN-CH_2-CO-$; both of these groups being named glycyl, this gives the name glycylglycylalanine. A hyphen indicates a C-l-to-N-2 peptide bond. Glutamic acid can be bound through either or both of its two carboxy groups, and Greek letters α and γ are used to indicate the position of the link.

Example

$$HOOC-CH_2-CH_2-CH(NH_2)-CO-$$

9.

5	4	3	2	1
	γ	β	α	

α-glutamyl

The dipeptide *N*-α-glutamylglycine is abbreviated as Glu-Gly. *N*-γ-glutamyl-▶ glycine is represented by the bond symbol ⌊ or ⌈⌋ and this kind of presentation is further illustrated below.

Examples

N-γ-glutamylglycine is

10.

Glu
⌊Gly   **or**   ⌈Gly
              Glu   **or**   Glu ⌈⌋Gly   **or**   Glu(-Gly)

Glutathione is

11.

Glu
⌊Cys-Gly   **or**   Glu ⌈⌋Cys-Gly   **or**   Glu(-Cys-Gly)

Great care should be exercised in presenting these formulae, because a simple vertical line between Glu and Cys would correspond to a thioester between the γ carboxy group of Glu and the SH group of Cys.

Example

12.

Glu
|
Cys–Gly

Cyclic peptides in which the ring consists solely of amino acid residues with eupeptide links may be called **homodetic** cyclic peptides. Three representations are possible. Gramicidin S is given as an example. In this decapeptide, all amino acids are L, with the exception of Phe which is D, as is shown by D-Phe or DPhe.

Examples

13.   cyclo(-Val-Orn-Leu-D-Phe-Pro-Val-Orn-Leu-D-Phe-Pro-)

14. 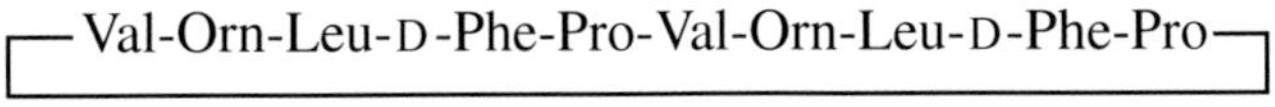   ⌐ Val-Orn-Leu-D-Phe-Pro-Val-Orn-Leu-D-Phe-Pro ⌐

15. 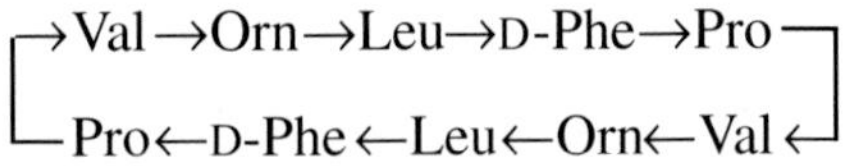   ⌐→Val→Orn→Leu→D-Phe→Pro ⌐
      └─Pro←D-Phe←Leu←Orn←Val ←┘

Heterodetic cyclic peptides are peptides consisting only of amino acid residues, but the links forming the ring are not solely eupeptide bonds; one or more is an isopeptide, disulfide, ester, *etc.*

Examples

16. 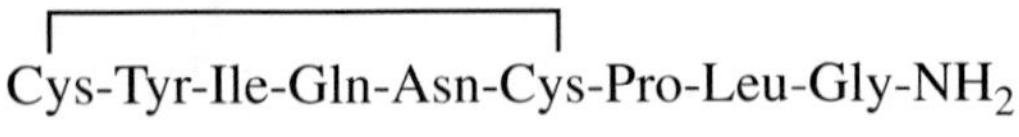   ┌─────────────────┐
      Cys-Tyr-Ile-Gln-Asn-Cys-Pro-Leu-Gly-NH$_2$

      Oxytocin

17. 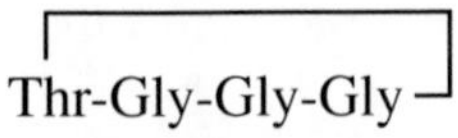   ┌──────────────┐
      Thr-Gly-Gly-Gly ┘

      Cyclic ester of threonylglycylglycylglycine

11.4   LIPID NOMENCLATURE

Lipids are substances of biological origin that are soluble in non-polar solvents. There are saponifiable lipids, such as acylglycerols (fats and oils), and waxes and phospholipids, as well as non-saponifiable compounds, principally steroids.

The term '**fatty acid**' designates any of the aliphatic carboxylic acids that can be liberated by hydrolysis from naturally occurring fats and oils. 'Higher fatty acids' are those that contain ten or more carbon atoms. Neutral fats are mono-, di-, or tri-esters of glycerol with fatty acids, and are therefore termed monoacylglycerol, diacylglycerol and triacylglycerol. Trivial names are retained for fatty acids and their acyl groups: stearic acid, stearoyl; oleic acid, oleoyl. Esters from glycerol are usually named by adding the name of the acyl group to that of glycerol.

Examples

1. 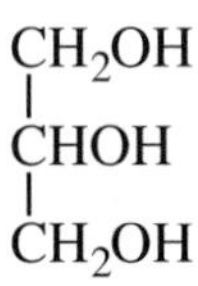 CH$_2$OH
    |
    CHOH
    |
    CH$_2$OH
    glycerol

2. 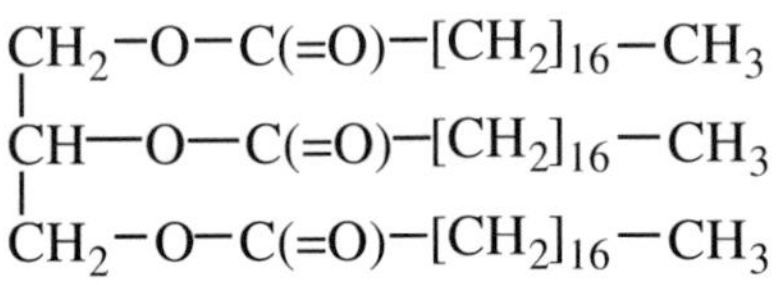 CH$_2$-O-C(=O)-[CH$_2$]$_{16}$-CH$_3$
    |
    CH-O-C(=O)-[CH$_2$]$_{16}$-CH$_3$
    |
    CH$_2$-O-C(=O)-[CH$_2$]$_{16}$-CH$_3$
    tristearoylglycerol

3. 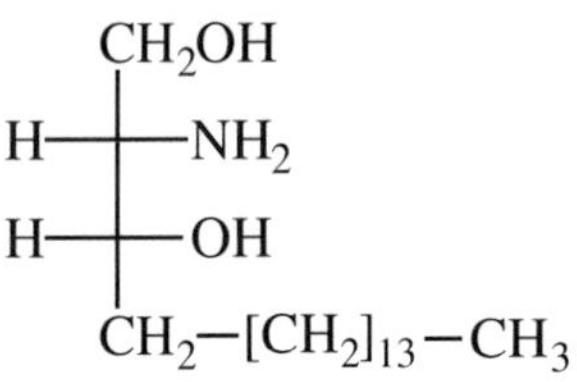 CH$_2$OH
    H—|—NH$_2$
    H—|—OH
    CH$_2$-[CH$_2$]$_{13}$-CH$_3$
    sphinganine

Compounds similar to glycerol, called sphingoids, are derivatives of sphinganine (D-*erythro*-2-aminooctadecane-1,3-diol). The trivial name sphinganine implies the configuration. Note the use of the stereodescriptor D-*erythro* in the systematic name.

Phospholipids are lipids containing phosphoric acid as a mono- or di-ester. When glycerol is esterified on C-1 by a molecule of phosphoric acid, the result is a chiral glycerol phosphate. A specific numbering is necessary to name it without ambiguity. The symbol *sn* (for stereospecifically numbered) is used: thus, D-glycerol and L-glycerol phosphates are easily recognised, the *sn*-glycerol 1-phosphate (Example 5) being the enantiomer of *sn*-glycerol 3-phosphate (Example 4).

Examples

4.
$$\begin{array}{c} CH_2OH \\ HO{\blacktriangleright}C{\blacktriangleleft}H \\ CH_2{-}O{-}PO_3H_2 \end{array}$$

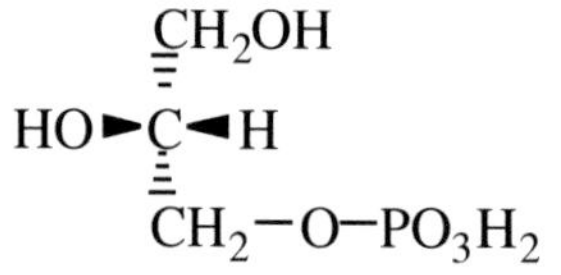

sn-glycerol 3-phosphate

5.
$$\begin{array}{c} CH_2{-}O{-}PO_3H_2 \\ HO{\blacktriangleright}C{\blacktriangleleft}H \\ CH_2OH \end{array}$$

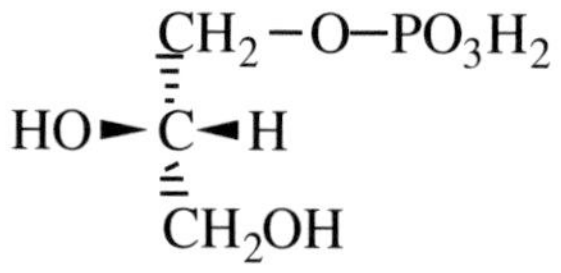

sn-glycerol 1-phosphate

A complete name is given as shown in Example 6 below for a phospholipid. Note that, exceptionally, the name ethanolamine in place of the systematic name 2-aminoethan-1-ol is widely used in the biochemical literature.

Example

6.

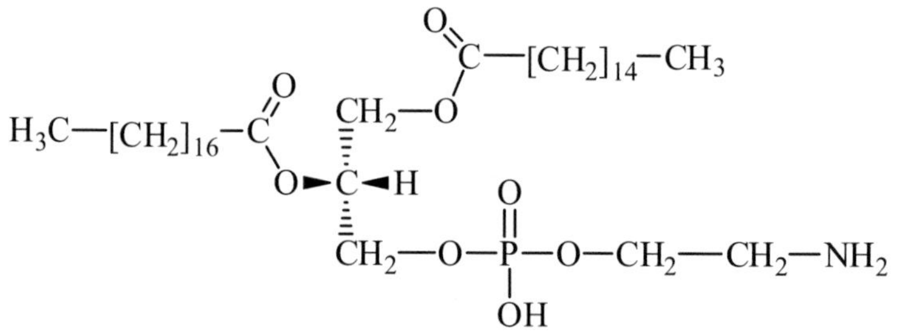

$$H_3C{-}[CH_2]_{16}{-}\underset{O}{\overset{O}{C}} \quad \underset{O{\blacktriangleright}C{\blacktriangleleft}H}{\overset{CH_2{-}O{-}\overset{O}{C}{-}[CH_2]_{14}{-}CH_3}{}} $$

$$CH_2{-}O{-}\underset{OH}{\overset{O}{P}}{-}O{-}CH_2{-}CH_2{-}NH_2$$

1-palmitoyl-2-stearoyl-*sn*-glycero-3-phosphoethanolamine

11.5   STEROID NOMENCLATURE

Steroids are compounds possessing the tetracyclic skeleton of cyclopenta[*a*]▶ phenanthrene (1) below, or a skeleton formally derived from it by one or more bond scissions, or ring expansions, or contractions. Natural steroids have trivial names. The nomenclature of steroids is not based on these trivial names, but on a few parent structures that are common to many compounds. Some of these are shown below. They represent the absolute configurations and are numbered as shown in (2).

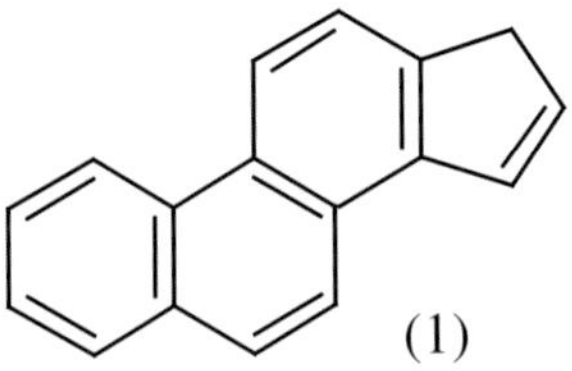

(1) 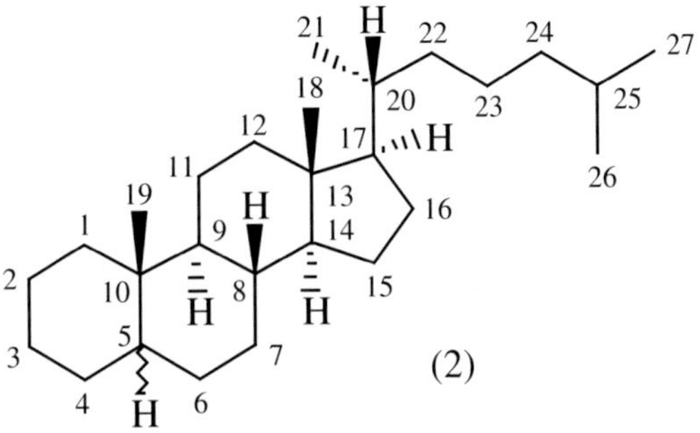   (2)

Substitutive nomenclature is used to designate characteristic groups and unsaturation. Structural modifications are expressed by appropriate non-detachable prefixes. According to the projection formula, substituents are designated α (below the plane of the ring) or β (above that plane). When no indication is given, the structural skeleton and the configurations at carbon atoms 8, 9, 10, 13, 14, 17 and 20 are as shown in (2). The configuration at C-5 is either 5α or 5β; it must always be stated in the name. Any unknown configuration is denoted by ξ (Greek xi). In structural formulae, broken, thick, and wavy bonds (⌇⌇⌇) correspond to the stereodescriptors α, β, and ξ, respectively. Examples 1–8 show some trivial and systematic names.

Examples

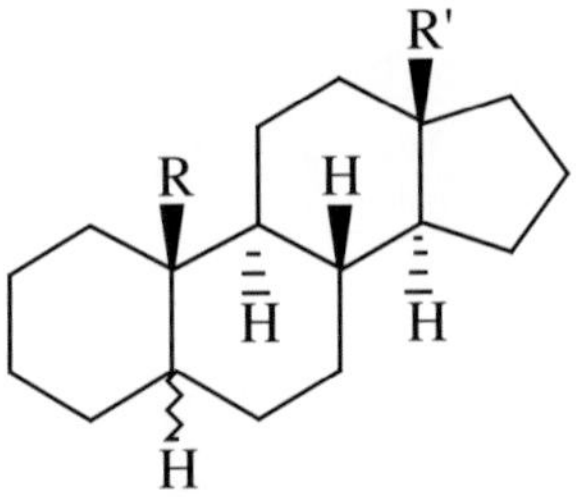

1. (a) estrane R = H; R' = CH_3
   (b) androstane R = R' = CH_3

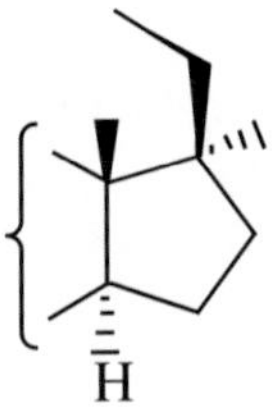

2. pregnane R = CH_3

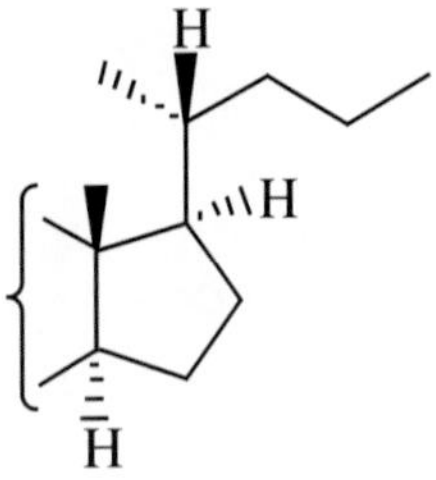

3. cholane R = CH_3

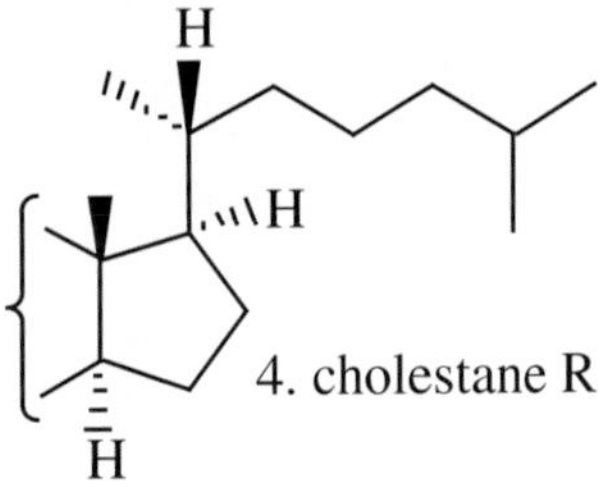

4. cholestane R = CH_3

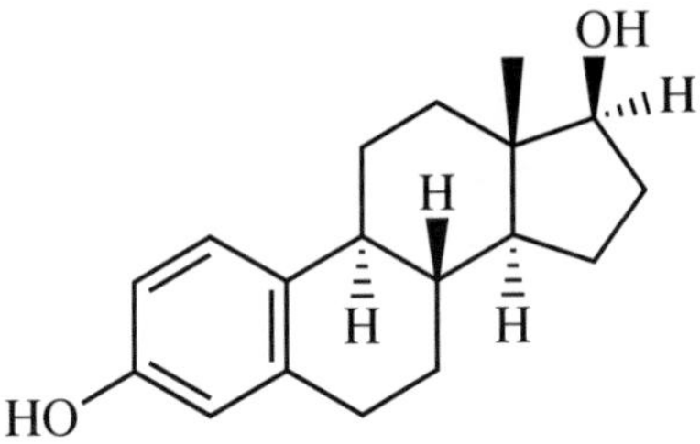

5. estradiol
estra-1,3,5(10)-trien-3,17β-diol

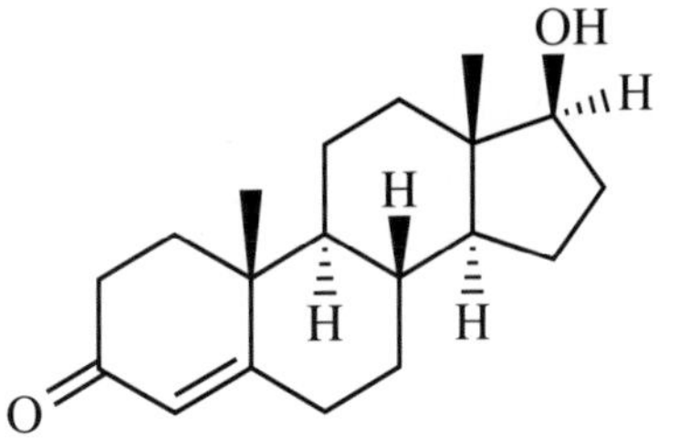

6. testosterone
17β-hydroxyandrost-4-en-3-one

167

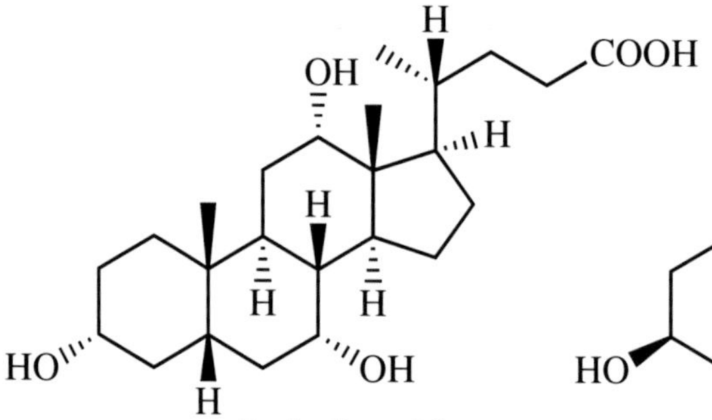

7. cholic acid
3α,7α,12α-trihydroxy-5β-cholan-24-oic acid

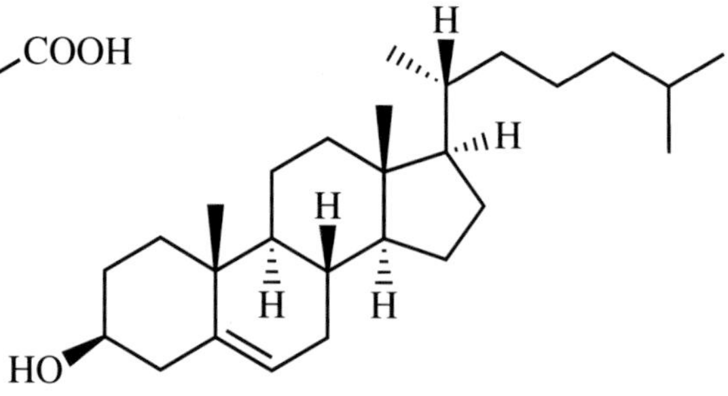

8. cholesterol
cholest-5-en-3β-ol

Non-detachable prefixes are used to indicate modifications affecting rings: 'homo' (enlarging), 'seco' (opening) and 'nor' (contracting). Locants are supplied as necessary to indicate the positions of modification.

Examples

9.

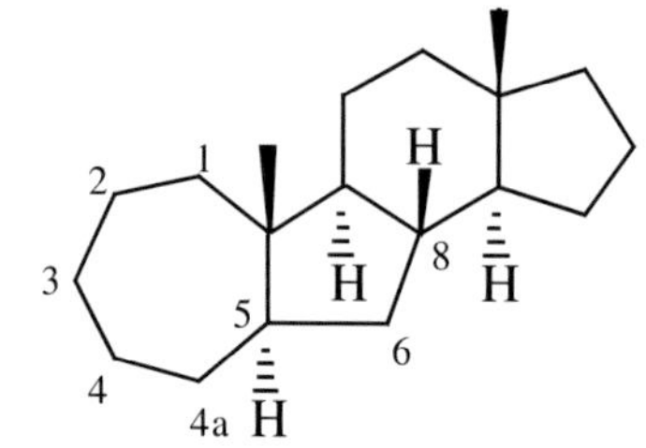

4a-homo-7-nor-5α-androstane

10.

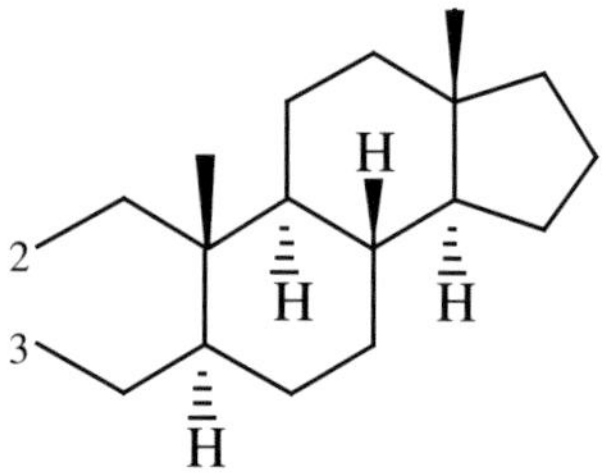

2,3-seco-5α-androstane

Much of steroid nomenclature has been applied to many other groups of natural products where there is a standard parent, *i.e.*, a parent with fixed configuration and orientation. This includes the use of α/β for configuration and the use of 'homo' and 'nor'.

11.6   NUCLEIC ACIDS

There are two types of **nucleic acid**, deoxyribonucleic acids (DNA) and ribonucleic acids (RNA). They are composed of long chains of phosphoric diesters with a phosphate group esterified to C-3 of one (deoxy)ribose monosaccharide unit and to C-5 of the next (deoxy)ribose unit (see Section 11.2 on carbohydrate nomenclature). At C-1 of the (deoxy)ribose there is attached a heterocyclic base. With DNA the base may be the pyrimidines cytosine or thymine, or the purines adenine or guanine. In RNA thymine is replaced by uracil. The (deoxy)ribose–base unit is termed a **nucleoside** (or more precisely a ribonucleoside or a deoxyribonucleoside). The locants for the atoms of the heterocyclic base are not primed whereas those of the ribose are primed. With a phosphate group attached at C-3 or C-5 of the ribose it is a **nucleotide** (or ribonucleotide or deoxyribonucleotide). In DNA and RNA much of the structure exists as a double helix composed of two separate chains (or two portions of the same chain). The two chains are held together with hydrogen bonds between the bases guanine and cytosine, and adenine pairing with thymine or uracil. Short chains of nucleotides are known as **oligonucleotides**.

Examples

1.

cytidine R = OH
2′-deoxycytidine R = H

2.

uridine R = OH, R′ = H
2′-deoxyuridine R = R′ = H
ribosylthymine R = OH, R′ = CH$_3$
thymidine R = H, R′ = CH$_3$

3.

adenosine R = OH
2′-deoxyadenosine R = H

4.

guanosine R = OH
2′-deoxyguanosine R = H

The nucleotide units of an oligonucleotide are treated as the nucleoside 3′-phosphate. Based on the trivial name 3′-adenylic acid for adenosine 3′-phosphate, the residue is called adenylyl. Other nucleotides are treated similarly (see Table 11.2). The links between each nucleotide need to be specified, *e.g.*, (3′ → 5′) (see Examples 7 and 8 below).

There are two accepted systems of abbreviation for trivial names, using either one or three letters (see Table 11.2). With the three letter system the phosphate group is represented by an italic *P*. When the *P* is on the left the phosphate is attached at 5′ and when it is on the right it is attached at 3′. Deoxynucleosides are indicated by a d before the relevant symbol.

Examples
   5. thymidine = dThd
   6. adenosine 5′-phosphate = *P*Ado = AMP (adenosine monophosphate)

Table 11.2 Names and their abbreviations and symbols for nucleic acid components

Base		Ribonucleoside		RNA nucleotide	
Trivial name	Symbol	Trivial name	Symbol	Trivial name	Symbol
adenine	Ade	adenosine	Ado	adenylyl	A
cytosine	Cyt	cytidine	Cyd	cytidylyl	C
guanine	Gua	guanosine	Guo	guanylyl	G
thymine	Thy	ribosylthymine	Thd	uridylyl	U
uracil	Ura	uridine	Urd		

With the one-letter system, a terminal phosphate is represented by the letter p (or pp for a diphosphate, or ppp for a triphosphate, *etc.*). It is assumed that the phosphate diester link is $3' \rightarrow 5'$. With a deoxyoligonucleotide the d is placed before the sequence in brackets.

Examples

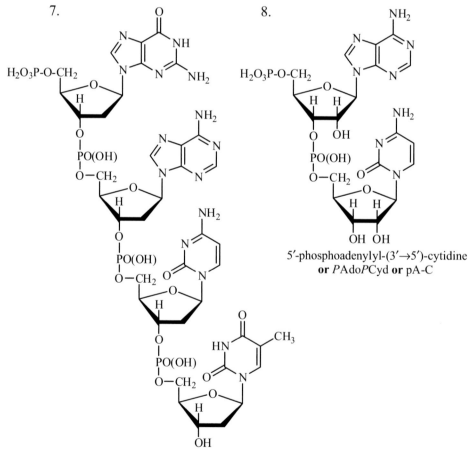

5'-phosphoadenylyl-$(3' \rightarrow 5')$-cytidine
or *P*Ado*P*Cyd **or** pA-C

5-phospho-2'-deoxyguanylyl-$(3' \rightarrow 5')$-phospho-2'-deoxyadenylyl-$(3' \rightarrow 5')$-►
2'-deoxycytidylyl-$(3' \rightarrow 5')$-thymidine
or *P*dGuo-3'*P*5'-dAdo-3'*P*5'-dCyd-3'*P*5'-dThd **or** d(pG-A-C-T)

11.7   VITAMINS

The names and structures of the principal **vitamins** are summarised here. Vitamins are food products which are required for the normal functioning of the body. Their absence or deficiency causes specific deficiency diseases. Vitamin A is a common name for the fat soluble vitamin retinol, Example 1(a). It is a polyene which, when oxidised, gives the visual pigment retinal, Example 1(b).

Example

1.

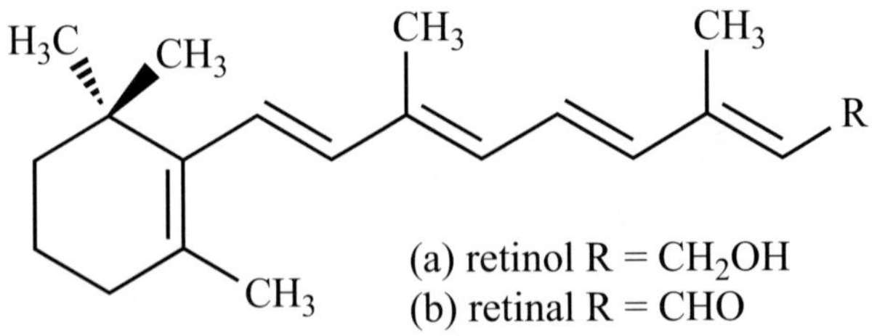

(a) retinol R = CH_2OH
(b) retinal R = CHO

The vitamin B complex is composed of four water soluble vitamins, B_1, B_2, B_6 and B_{12}. B_1 is the common name for thiamine, Example 2. Its phosphoric ester is an important coenzyme for oxidative decarboxylation and a number of other enzyme catalysed reactions. B_2 is the common name for riboflavin, Example 3. It forms part of the flavin nucleotide coenzymes FAD (flavoadenine) and FMN (flavin mononucleotide) involved in a number of redox enzyme reactions.

Examples

2.

thiamine

3.

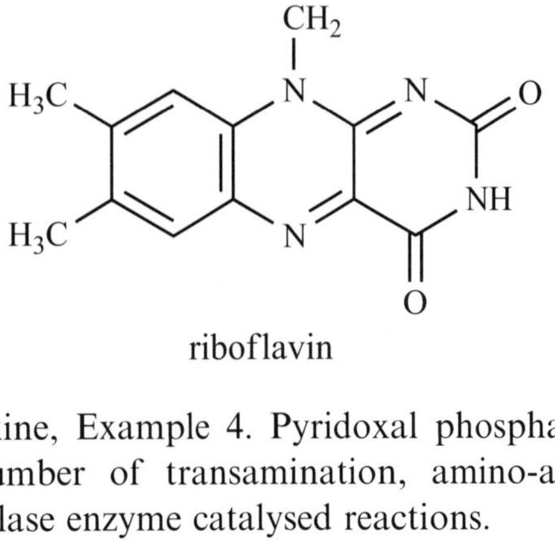

riboflavin

B_6 is the common name of pyridoxine, Example 4. Pyridoxal phosphate, Example 5, is a coenzyme for a number of transamination, amino-acid decarboxylation, racemisation and aldolase enzyme catalysed reactions.

171

Examples

4. 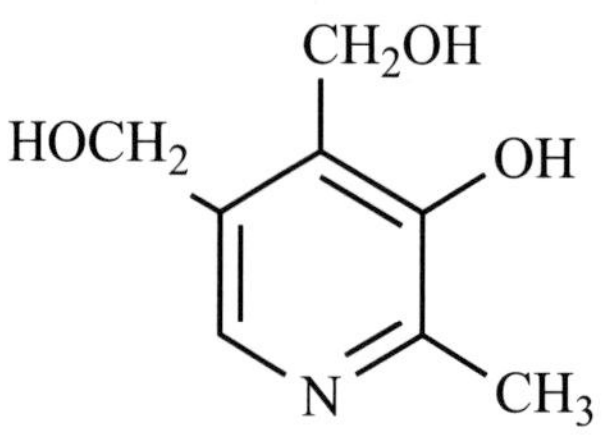 pyridoxine

5. 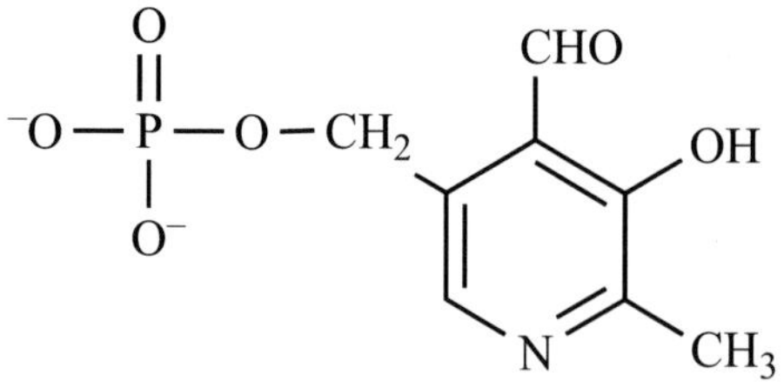 pyridoxal phosphate

B_{12} is the common name for cobalamin, Example 6. The name is generally used to imply cyanocobalamin although there are many other closely related compounds. Methylcobalamin is a coenzyme.

Examples

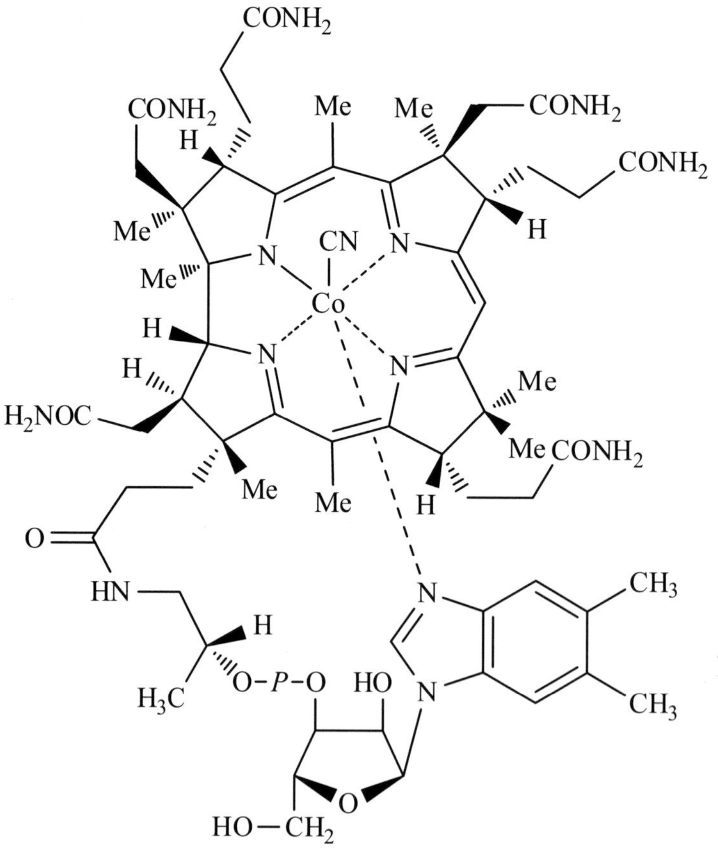

6. cyanocobalamin

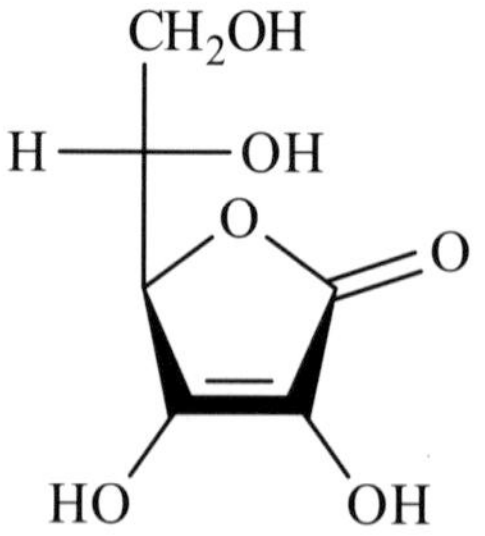

7. ascorbic acid

Vitamin C is the common name of the water soluble vitamin ascorbic acid, Example 7. Its deficiency causes scurvy. Vitamin D is the common name for a group of secosteroids. The natural vitamin D, photochemically produced from cholesta-5,7-dien-3β-ol, Example 8(a). Ergocalciferol, Example 8(b), readily prepared photochemically from the yeast steroid ergosterol, often replaces it in foodstuffs.

Examples

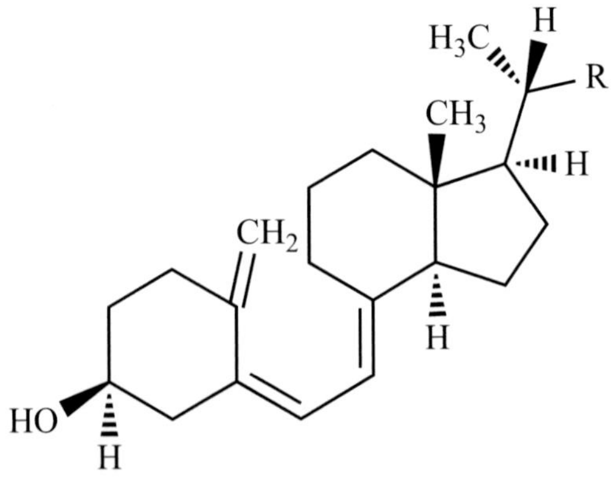

General formula of vitamin D

8(a). cholecalciferol, R =

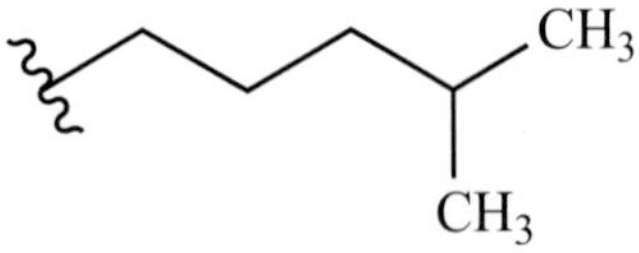

8(b). ergocalciferol, R =

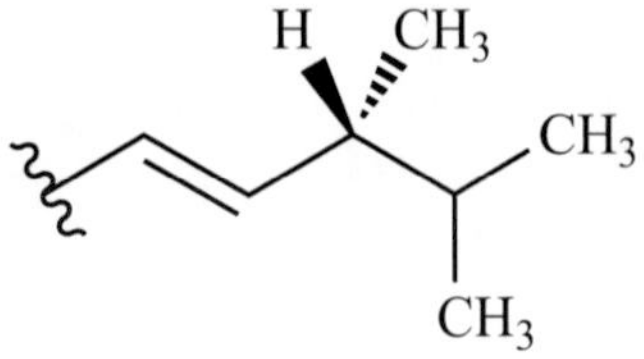

Vitamin E is the common name of the fat soluble antioxidant α-tocopherol, Example 9.

Example

9.

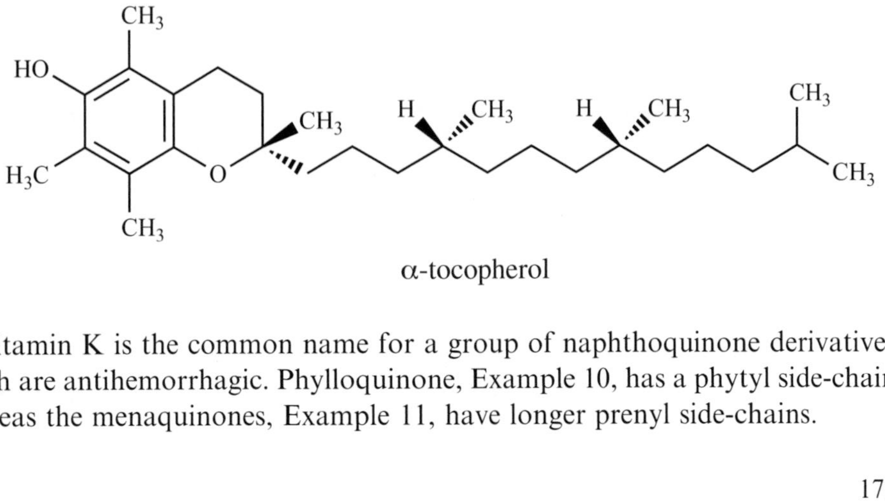

α-tocopherol

Vitamin K is the common name for a group of naphthoquinone derivatives which are antihemorrhagic. Phylloquinone, Example 10, has a phytyl side-chain whereas the menaquinones, Example 11, have longer prenyl side-chains.

Examples

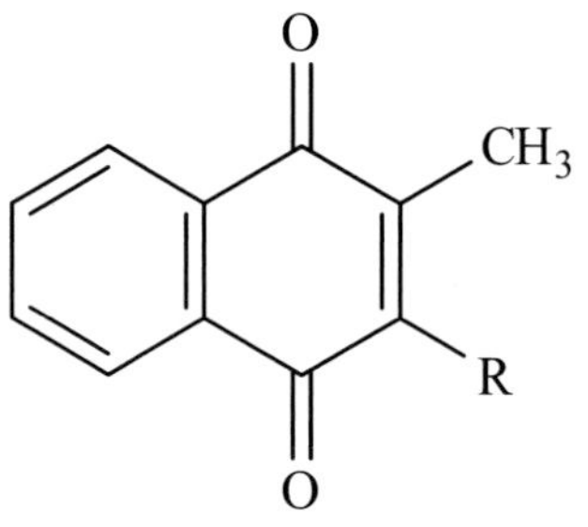

General formula of vitamin K

10. phylloquinone R =

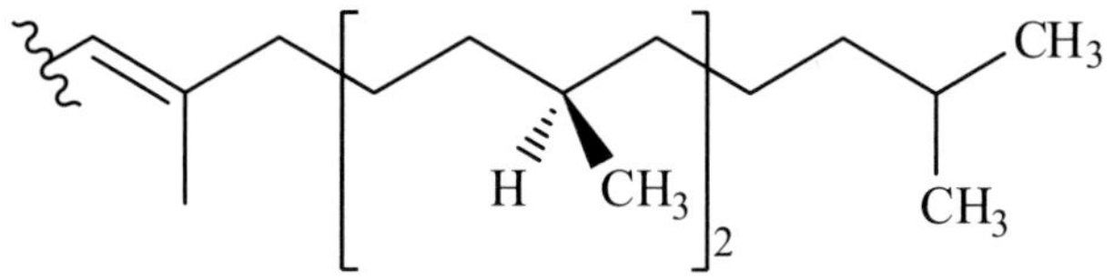

11. menaquinone R =

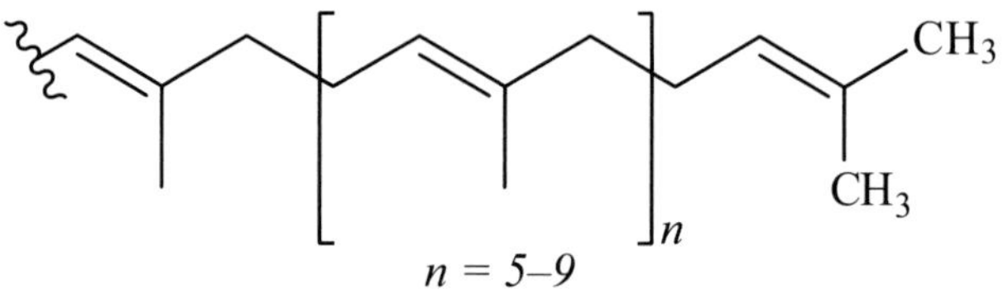

$n = 5–9$

FURTHER READING AND REFERENCES

Biochemical Nomenclature and Related Documents, 2nd edition, Portland Press, 1992. For Web versions of most of the contents see http://www.chem.qmul.ac.uk/iupac/bibliog/white.html

Revised Section F: Natural products and related compounds. IUPAC Recommendations 1999, *Pure and Applied Chemistry*, 1999, **71**, 587–643. For a Web version see http://www.chem.qmul.ac.uk/iupac/sectionF and for a PDF see http://www.iupac.org/reports/1999/7104giles/index.html

Nomenclature of Carbohydrates. Recommendations 1996, *Pure and Applied Chemistry*, 1996, **68**, 1919–2008. For a Web version see http://www.chem.qmul.ac.uk/iupac/2carb/ and for a PDF see http://www.iupac.org/publications/pac/1996/pdf/6810x1919.pdf (3895 kB).

Nomenclature and symbolism for amino acids and peptides. Recommendations 1983, *Pure and Applied Chemistry*, 1984, **56**, 595–624. For a Web version see http://www.chem.qmul.ac.uk/iupac/AminoAcid/ and for a PDF see http://www.iupac.org/publications/pac/1984/pdf/5605x0595.pdf (527 kB).

Nomenclature of lipids (1976), *European Journal of Biochemistry*, 1977, **79**, 11–21. For a Web version see http://www.chem.qmul.ac.uk/iupac/lipid/

Nomenclature of steroids, Recommendations 1989, *Pure and Applied Chemistry*, 1989, **61**, 1783–1822. For a web version see http://www.chem.qmul.ac.uk/iupac/steroid/ and for a PDF see http://www.iupac.org/publications/pac/1989/pdf/6110x1783.pdf (1597 kB).

Abbreviations and symbols for nucleic acids, polynucleotides and their constituents (1970), *Pure and Applied Chemistry*, 1974, **40**, 277–290. For a Web version see http://www.chem.qmul.ac.uk/iupac/misc/naabb.html and for a PDF see http://www.iupac.org/publications/pac/1974/pdf/4003x0277.pdf (385 kB).

Nomenclature of retinoids. Recommendations 1981, *European Journal of Biochemistry*, 1982, **129**, 1–5; *Pure and Applied Chemistry*, 1983, **55**, 721–726. For a Web version see http://www.chem.qmul.ac.uk/iupac/misc/ret.html and for a PDF see http://www.iupac.org/publications/pac/1983/pdf/5504x0721.pdf (128 kB).

Nomenclature for vitamins B-6 and related compounds (1973), *European Journal of Biochemistry*, 1973, **40**, 325–327; *Pure and Applied Chemistry*, 1973, **33**, 445–452. For a Web version see http://www.chem.qmul.ac.uk/iupac/misc/B6.html and for a PDF see http://www.iupac.org/publications/pac/1973/pdf/3302x0445.pdf (211 kB).

Nomenclature of corrinoids (1973), *European Journal of Biochemistry*, 1974, **45**, 7–12; *Pure and Applied Chemistry*, 1976, **48**, 495–502. For a Web version see http://www.chem.qmul.ac.uk/iupac/misc/B12.html and for a PDF see http://www.iupac.org/publications/pac/1976/pdf/4804x0495.pdf (156 kB).

Nomenclature of vitamin D. Recommendations 1981, *European Journal of Biochemistry*, 1982, **124**, 223–227; *Pure and Applied Chemistry*, 1982, **54**, 1511–1516. For a Web version see http://www.chem.qmul.ac.uk/iupac/misc/D.html and for a PDF see http://www.iupac.org/publications/pac/1982/pdf/5408x1511.pdf (180 kB).

Nomenclature of tocopherols and related compounds. Recommendations 1981, *European Journal of Biochemistry*, 1982, **123**, 473–475; *Pure and Applied Chemistry*, 1982, **54**, 1507–1510. For a web version see http://www.chem.qmul.ac.uk/iupac/misc/toc.html and for a PDF see http://www.iupac.org/publications/pac/1982/pdf/5408x1507.pdf (115 kB).

Nomenclature of quinones with isoprenoid side chains (1973), *European Journal of Biochemistry*, 1975, **53**, 15–18; *Pure and Applied Chemistry* 1974, **38**, 439–447. For a Web version see http://www.chem.qmul.ac.uk/iupac/misc/quinone.html and for a PDF see http://www.iupac.org/publications/pac/1975/pdf/3803x0439.pdf (416 kB).

12 Other Types of Nomenclature

12.1 ## OTHER SYSTEMS OF NOMENCLATURE USED FOR CHEMICAL COMPOUNDS

IUPAC is the major provider of nomenclature for practicing chemists, but since not only chemists work with chemical compounds other groups of scientists have developed their own kinds of nomenclature, which are particularly suited to their own specific needs. This Chapter attempts to deal with some of the most important from a chemist's point of view. However, mineral names and solid-state nomenclature, for example, have been omitted, as being at present too far outside the scope of current IUPAC recommendations. There is much nomenclature information available on the Web, but readers should be wary of adopting what they find unless they can be sure that it originated from a reliable source, and that it is sanctioned by the appropriate authorities.

While IUPAC remains the ultimate source of international chemical nomenclature, these other nomenclatures are used in more restricted contexts. The student will often come across them, and should be aware of their existence and their function. Subjects such as enzymes, pesticides, and cosmetics are discussed below. Pharmaceuticals also have their internationally recognised names.

12.2 CHEMICAL ABSTRACTS SERVICE NOMENCLATURE

The Chemical Abstracts Service (CAS), a Division of the American Chemical Society, provides a searchable database for subjects, authors, and compounds, *etc.* It has its own system of nomenclature. This initially was developed in close parallel and contact with IUPAC, but because IUPAC works rather slowly, whereas CAS aims to publish abstracts of all chemistry literature as soon as possible after it first appears, the two systems have diverged somewhat. The general pattern of CAS nomenclature is rather similar to IUPAC nomenclature. However, CAS nomenclature is not identical to IUPAC nomenclature, and users of either system should be aware that this is the case.

A useful aspect of Chemical Abstracts Service nomenclature is the **Registry Number**. A registry number is assigned to every compound as soon as it is entered into the CAS database. It is intended to be a unique identifier of the compound to which it is assigned, but it is simply a number and carries no information whatsoever concerning the name or structure of that compound. It may be invaluable to anyone using it to interrogate the CAS database, and it is sometimes quoted in publications. However, some compounds now have more

Principles of Chemical Nomenclature: A Guide to IUPAC Recommendations
Edited by G. J. Leigh
© International Union of Pure and Applied Chemistry 2011
Published by the Royal Society of Chemistry, www.rsc.org

than one registry number, because they were entered into the database under a different name, and some registry numbers cover more than one compound. As an identifier it performs part of the function of the IUPAC InChI described in Chapter 14, but the latter also gives information about the compound structure, which a registry number can never do.

12.3 IUBMB ENZYME NOMENCLATURE

Classification of **enzymes** is standardised by the International Union of Biochemistry and Molecular Biology (IUBMB), and is based on the reaction which a given enzyme catalyses. Six main groups of reaction are recognised, labelled EC 1 to EC 6, where EC stands for Enzyme Commission.

EC 1 Oxidoreductases, which catalyse oxidation and reduction reactions.
EC 2 Transferases, which catalyse the transfer of an alkyl, acyl, glycosyl, or other group, from one substrate to another.
EC 3 Hydrolases, which catalyse hydrolysis reactions.
EC 4 Lyases, which catalyse other cleavage reactions.
EC 5 Isomerases, which catalyse isomerisation reactions.
EC 6 Ligases, which catalyse bond-forming reactions for which ATP is required.

Each class is subdivided, depending on the compounds involved. For example, EC 1.1 is the oxidation of alcohols, EC 1.2 the oxidation of aldehydes, *etc.* These subclasses are in turn usually subdivided. EC 1.1.1 uses NAD(P)H, EC 1.1.2 uses cytochrome, EC 1.2.3 uses oxygen, *etc.* Each individual enzyme is then given a fourth number to distinguish it from other members of the sub-subclass. These numbers are assigned in order as new members of the sub-subclass are identified. For example EC 1.1.1.1 is alcohol dehydrogenase.

The last printed list of enzymes appeared in 1992. Since then the list has been maintained and updated on-line (reference 1).

12.4 WORLD HEALTH ORGANISATION INTERNATIONAL NONPROPRIETARY NAMES (INNs)

To prevent confusion between the trade names of products from different companies and/or different countries the World Health Organisation (WHO) assigns an international nonproprietary name (INN) to each new **drug**. Originally, these were abbreviated forms of a chemical name because the INN needs to be short. Today, the suffix (the 'stem' in WHO terminology) used reflects the pharmacological and/or chemical category of the compound or (potential) drug. For example atenolol has the suffix -olol which indicates a β-adrenoreceptor antagonist (beta-blocker). The names are, in the first instance, given in Latin, English, French and Spanish. Russian, Chinese and Arabic forms are also available. The main part of the name is designed to be distinct from all other INNs and other names, both when pronounced and when written. Care is taken to ensure that the name is not associated with an inappropriate meaning in other languages. In order to be usable in all languages WHO also has to restrict which letters are used. The letters h and k are omitted; the diphthongs ae and oe are rendered as e; the letter i is always used where y might originally have been selected and the combinations of letters th and ph are always rendered t and f, respectively. Where appropriate, INNs usually represent the free acid or base, so that salts and other derivatives are named with an appropriate addition, *e.g.*, oxacillin sodium is the name used for the sodium salt of oxacillin. So far about 9000 INNs have been assigned. The most recent published list of names may be found at http://www.who.int/medicines/services/inn/stembook/en/index.html.

12.5 ISO-APPROVED COMMON NAMES FOR PESTICIDES

Pesticides are substances or manufactured products that are intended to kill, repel, or otherwise control any organism that is designated a pest, and include herbicides (weedkillers), rodenticides, fungicides, bactericides, *etc.* Like most medicinal drugs, many pesticides are substances containing molecules of quite complex structure, and the systematic names of these compounds often become long and impractical for commercial and regulatory purposes.

A way of providing short and simple semi-systematic names for pesticides has been devised. These are based upon several standards of the International Organisation for Standardisation (ISO) (references 2–4). Although ISO and IUPAC collaborate, ISO names are not necessarily consistent with IUPAC recommendations. Administration of the system and approval of new names is controlled by a standing technical committee, ISO/TC 81, with the title *Common names for pesticides and other agrochemicals*.

The resulting non-proprietary names resemble the drug INNs granted by WHO (see Section 12.4 above). Examples are chlorpyrifos, iminoctadine, paraquat, and parathion. There is a list of recommended syllables that signalise specific chemical structures such as 'azin' for chlorine-substituted 1,3,5-triazines (*e.g.*, in atrazin) or 'nil' for nitriles (*e.g.*, in chloroxynil). In a few cases acronyms, which may even include numeric locants, are accepted, due to long-time usage. Examples include 2,4-D and DDT. When it is necessary to indicate that the compound is a salt or an ester, pesticide names may contain parts of normal chemical names, *e.g.,* fosetyl-aluminium, chlorpyrifos-methyl, dinoseb acetate, and paraquat dichloride. In this context, a number of substituent groups and ions have recommended names that are to be used in common names, as in iminoctadine trialbesilate. For further examples, see Table 12.1.

The texts of ISO standards are not available without payment and must be purchased, but a listing of names for pesticides is accessible on the internet in the *Compendium of Pesticide Common Names* (reference 5).

12.6 INTERNATIONAL NOMENCLATURE OF COSMETIC INGREDIENT NAMES (INCI NAMES)

For trade and regulatory purposes, a special nomenclature system has been developed for ingredients of **cosmetics** products. It is called INCI (International Nomenclature of Cosmetic Ingredients). The INCI names and the rules for

Table 12.1 Examples of names of ions and substituent groups to be used in ISO common names of pesticides

ISO name	IUPAC systematic name
albesilate	alkylbenzenesulfonate
butometyl	2-butoxypropan-2-yl
butotyl	2-butoxyethyl
diclexin	dicyclohexylammonium
dimolamine	(2-hydroxyethyl)dimethylammonium
diolamine	bis(2-hydroxyethyl)ammonium
ethadyl	ethane-1,2-diyl **or** ethylene
etolyl	2-ethoxyethyl
meptyl	octan-2-yl **or** 1-methylheptyl
metilsulfate	methyl sulfate
mexyl	heptan-2-yl **or** 1-methylhexyl
olamine	(2-hydroxyethyl)ammonium
tefuryl	(oxolan-2-yl)methyl **or** (tetrahydrofuran-2-yl)methyl
trimesium	trimethylsulfanium
trolamine	tris(2-hydroxyethyl)ammonium

forming them may be accessed on-line (references 6 and 7). These sources are provided by the European Union (EU), which increasingly makes use of this nomenclature system in rules for specification of ingredients, not only in cosmetics but also in personal care products in general and in household detergents. New names must be applied for at the *Personal Care Products Council*, formerly the *Cosmetics, Toiletries and Fragrance Association*, CTFA.

INCI names range from names that are equivalent to IUPAC names for particular well-defined substances to names that designate preparations of complex compositions obtained from microbial, plant or animal sources. Thus, a name such as butoxydiglycol is just a trivial name that is retained within the INCI system [for 2-(2-butoxyethoxy)ethan-1-ol], while more complex examples include ceteth-24, which is the INCI name for a mixture of α-hexadecyl-ω-hydroxypoly(oxyethane-1,2-diyl) molecules with an average number of 24 oxyethane-1,2-diyl units. 'Cet' is an abbreviation for cetyl, systematically hexadecyl, and 'eth' stands for ethoxylate. Coco-rapeseedate comprises coco alkyl esters of fatty acids from rapeseed oil. 'Coco' generally denotes mixtures of unbranched acyl and alkyl groups with 10 to 14 carbon atoms, as found in the fatty acids of the triacylglycerols of coconut oil and the alcohols derived from them.

REFERENCES

1. A current list of enzyme names can be found at http://www.chem.qmul.ac.uk/iubmb/enzyme/

2. International standard ISO 1750: 1981 (with later revisions and additions published twice yearly): *Pesticides and other agrochemicals – Common names*, International Organization for Standardization.

3. International Standard ISO 257: 2004 (E): *Pesticides and other agrochemicals – Principles for the selection of common names*, 3rd ed., 2004-06-15. International Organization for Standardization.

4. International Standard ISO 765: 1976: *Pesticides considered not to require common names*. International Organization for Standardization.

5. *Compendium of Pesticide Common Names*, http://www.alanwood.net/pesticides/index.html

6. See http://ec.europa.eu/consumers/sectors/cosmetics/cosing/ingredients, where an inventory of cosmetics ingredients with INCI names is available.

7. de Polo, K. F., *A short Textbook of Cosmetology*, Verlag für Chemische Industrie, H. Ziolkowski, Augsburg, 1998, may be of value as a guide.

13   Name Construction and Deconstruction

13.1   INTRODUCTION

Names and formulae must be of a form that can be interpreted by a reader, who should then be able to produce a structural diagram. Clearly this is related to the methods described in earlier Chapters for producing names and formulae from structures, with steps presented pictorially in flow charts. Deciphering a name or formula can be approached in a number of different ways. Here we describe general approaches to deciphering the different kinds of name that are met with in IUPAC nomenclature.

The first step is to establish whether the name or formula describes just one structure, or whether there is more than one structural unit present. For example, a compound may be made up of a cationic unit and an anionic unit (a binary name), and the structures of each of these units should then be derived separately from the name. A compound, of which a hydrated salt would be an example, may be made up of cations, anions, and neutral species, with the ratios between these species being fixed for the given compound, and defined in the name using multiplicative prefixes. Generally, cation names precede those of anions, and names of neutral species come last in the overall name of a compound. The names for these components are generally separated from each other in the overall compound name. Each component of the overall name, whether cationic, anionic, or neutral, should be deciphered separately.

Apart from general indexing, note that organic and inorganic chemistry rely on formulae in different ways. Organic chemists tend to draw out formulae so that the individual units of the compound are comprehensible to the informed chemist. Inorganic structures, and especially those of coordination compounds, are often written in line formulae that are not immediately recognisable. Hence some guidance is provided below on the interpretation of the formulae of coordination compounds.

The portion of a name relating to a cation can be recognised by the suffix 'ium', except for complex cations such as those found in coordination/organometallic compounds. In that case, the 'ium' suffix is not generally used. Instead, the charge on the ion may be given explicitly, or may need to be inferred from the oxidation number of the metal that should be provided in its stead. Coordination/organometallic compounds, whether they be cationic, anionic, or neutral, are normally easily identified through the indication of the presence of a metal atom or ion.

In general, the portion of a name that describes an anion can be recognised by the suffixes 'ate', 'ite', or 'ide'. The major exception to this is in the functional

Principles of Chemical Nomenclature: A Guide to IUPAC Recommendations
Edited by G. J. Leigh
© International Union of Pure and Applied Chemistry 2011
Published by the Royal Society of Chemistry, www.rsc.org

class nomenclature that is used to describe esters, acyl halides, and anhydrides. These can be recognised by the fact that the part of the compound name containing the suffix 'ate' or 'ide' is immediately preceded by the name of a neutral species, usually that of an organic fragment ending in 'yl'.

Once the name of a compound has been separated into its component parts, the next step is to identify the kind of nomenclature that is present in each. The presence of the term 'poly' within the name is an immediate indicator that the name probably represents a polymeric species. Guidance on deciphering the names of these species is provided in Chapter 9 and Section 13.4, below. Such names generally adapt the nomenclature for organic compounds (Chapter 6) and/or that of coordination compounds (Chapter 7). A strategy for deciphering the names of complexes is, in turn, detailed in Section 13.3, but because coordination compounds also often contain organic components as ligands, we begin first in Section 13.2 with strategies for deciphering the names of organic compounds, or of organic fragments within coordination complexes and polymers.

13.2   DECIPHERING NAMES OF ORGANIC COMPOUNDS

This section provides guidelines for deducing the structure of an organic compound from its systematic name. A systematic name in organic nomenclature generally consists of a parent hydride name that is modified by prefixes and suffixes. The three main parts of a name are: (i) prefix, or prefixes, corresponding to substituent groups; (ii) the name of a parent hydride which includes any non-detachable prefixes for heteroatoms, hydrogenation and skeletal modification or the name of a functional parent compound; and (iii) endings 'ene' and 'yne' indicating unsaturation, and suffixes, both characteristic and cumulative. These three parts have been discussed in Chapter 6 and are illustrated in the following scheme.

$$\textit{First} \begin{cases} \text{alphabetised} \\ \textbf{detachable} \\ \textbf{prefixes} \end{cases} \textit{then} \begin{cases} \text{the } \textbf{non-detachable} \\ \textbf{prefixes} \\ \text{hydro/dehydro} \end{cases} \textit{then} \begin{cases} \text{name of parent hydride} \\ \text{with any structural} \\ \textbf{non-detachable prefixes,} \\ \text{such as 'cyclo'} \end{cases}$$

$$\textit{then} \begin{cases} \textbf{endings} \\ \text{'ane' or} \\ \text{'ene'/'yne'} \end{cases} \textit{and, finally} \begin{cases} \text{all necessary} \\ \text{characteristic} \\ \textbf{suffixes} \end{cases}$$

To retrieve the structure from a name, a sequence of operations is followed to establish (a) the type of nomenclature (substitutive or functional class), (b) the parent hydride and its numbering, (c) the suffix (if any), (d) the detachable prefixes and 'ene'/'yne' endings. The guide to name construction cited in Section 6.10 will be of assistance in the opposite process, name decipherment.

Functional class nomenclature is normally used for esters, acid halides, and anhydrides. Anhydrides are easily recognised, as the term 'anhydride' is usually present in the name, replacing the term 'acid' in the name of the acid from which it is derived. Names of esters normally contain the name of an organic anion ending in 'ate' and are based upon the name of the acid from which it is derived, giving names such as 4-chlorohexyl acetate for $CH_3COOCH_2CH_2CH_2CHClCH_2CH_3$, methyl chloroacetate, for $ClCH_2COOCH_3$, and dimethyl sulfate for $SO_2(OCH_3)_2$. Acid halide names consist of the acid-derived fragment, usually with the suffix 'oyl' or 'yl', followed by the name of the halide, as in acetyl chloride for CH_3COCl.

The following Examples illustrate how to produce a structure from a substitutive name.

Example

1. 6-(5-hydroxypenta-1,3-diyn-1-yl)dodeca-2,4,7-triene-1,11-diol

The term 'dodeca' in this name indicates that this compound is derived from the straight-chain twelve-carbon saturated parent hydride dodecane.

$$CH_3-CH_2-CH_2-CH_2-CH_2-CH_2-CH_2-CH_2-CH_2-CH_2-CH_2-CH_3$$
$$12 \quad 11 \quad 10 \quad 9 \quad 8 \quad 7 \quad 6 \quad 5 \quad 4 \quad 3 \quad 2 \quad 1$$

The suffix 'ol' with its multiplicative prefix 'di' and locants '1' and '11' indicates the presence of two hydroxy groups (–OH) at positions 1 and 11 of dodecane chain. The ending 'ene' with its multiplicative prefix 'tri' and locant set 2,4,7 shows the presence of three double bonds at positions 2, 4 and 7 on the dodecane chain. The resulting parent structure with its numbering is:

$$CH_3-CH(OH)-CH_2-CH_2-CH=CH-CH_2-CH=CH-CH=CH-CH_2-OH$$
$$12 \quad 11 \quad\quad 10 \quad 9 \quad 8 \quad 7 \quad 6 \quad 5 \quad 4 \quad 3 \quad 2 \quad 1$$

The complex prefix (5-hydroxypenta-1,3-diyn-1-yl) with its locant 6 describes a substituent at position 6 of the parent structure. The substituent is a five-carbon chain derived from the parent hydride pentane. The prefix 'hydroxy' with its locant 5 denotes a hydroxy group in position 5; the infix 'yn' with its locant set 1,3 indicates two triple bonds at positions 1 and 3; the suffix 'yl' with its locant 1 shows that the parent has lost a hydrogen atom at position 1. The structure of this substituent is therefore as shown.

$$HO-CH_2-C\equiv C-C\equiv C-$$

The complete structure is presented below.

$$HO-CH_2-C\equiv C-C\equiv C$$
$$|$$
$$CH_3-CH(OH)-CH_2-CH_2-CH=CH-CH-CH=CH-CH=CH-CH_2-OH$$
$$12 \quad 11 \quad\quad 10 \quad 9 \quad 8 \quad 7 \quad 6 \quad 5 \quad 4 \quad 3 \quad 2 \quad 1$$

Example

2. 4-bromo-2-chlorobenzoic acid

Following the sequence of operations (a)–(d) listed on page 182, this is a substitutive name derived from the retained name benzoic acid (a). Applying (b) and (c) implies that this is a functional parent compound. The unexpressed locant '1' is assigned to the principal characteristic group –COOH and the allowed locant set is 1,2,4, lower than the alternative 1,4,6 (see structure immediately below). The substituent groups are thus present at positions 2 and 4; the bromo-substituent is assigned to position 4, and the chloro-substituent to position 2. The structure is as shown.

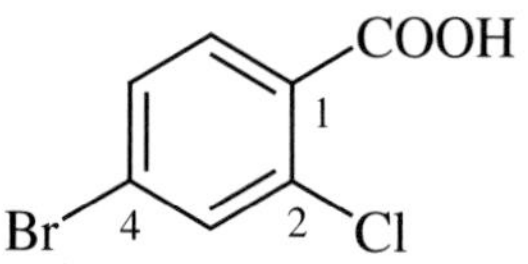

Had Example 2 been a functional class name, this would have been suggested by a two-part structure and a termination such as 'ate', indicating an ester, as in Example 3 below.

Example

3. methyl 6-methylanthracene-2-carboxylate

The part name 6-methylanthracene-2-carboxylate is a substitutive name in which the suffix 'carboxylate' indicates a group at position 2 of the parent hydride with the retained name, anthracene, and the numbering of which is shown in Section 6.3.4, Example 28. A methyl substituent is at position 6 of the parent hydride. The separate name methyl at the front of the substitutive name and associated with the suffix carboxylate at the end of the second part of the name means that an ester is present, represented by the grouping –CO–O–CH$_3$. The full structure is thus:

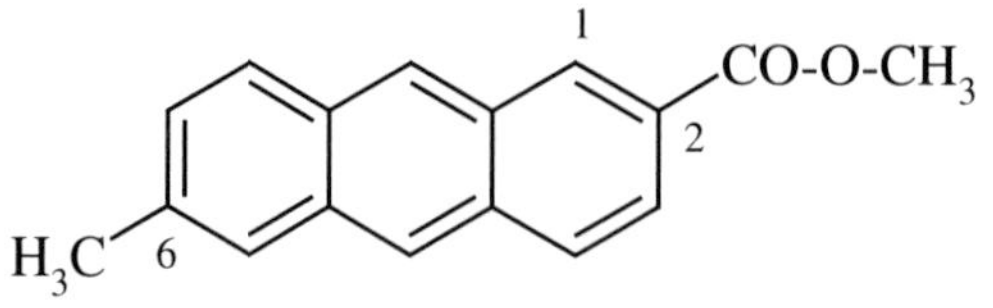

13.3 DECIPHERING NAMES AND FORMULAE OF COORDINATION COMPOUNDS

Coordination and organometallic complexes consist of one or more central atoms surrounded by ligands. Their names are constructed additively, which means that the first step in a deciphering process is usually to identify the central atom(s). Any metal atoms present are often central atoms. Generally the central atoms can be found at the end of the name or at the beginning of the formula. The major exception to this convention is where the creator of the formula may have wished to emphasise some specific information about the structure of the molecule. Commonly this occurs in dinuclear species. The remainder of the name or formula represents ligands. Systematic and trivial ligand names, formulae, and abbreviations must all be considered in order to decipher the ligand structures.

The coordination number and geometry of the central atom can be identified from the polyhedral symbol found in front of the name or formula (see Table P5). If there is no polyhedral symbol, the coordination number will have to be determined by identifying the number of coordinating atoms provided by the ligands. The geometry may have to be inferred from the context, or assumed through use of chemical knowledge, for example, from the observation that among six-coordinate complexes octahedral geometries are much more common than trigonal prismatic.

Once the ligands have been identified, the coordination sites can be assigned to the various ligands. κ-Terms indicate which atoms of the ligands are actually bonded to the central atom. A ligand with a μ prefix is a bridging ligand, and if there is a numerical subscript immediately following the μ, that subscript indicates the number of central atoms that the ligand bridges, greater than the minimum two. An η symbol preceding a ligand name indicates that contiguous atoms are bound to the central atom (see Section 8.2.5.1). The right superscript on the η indicates the number of contiguous ligand atoms bound to the central atom.

Care must be taken where a particular isomer is identified using a configuration index (see Section 3.9.2). The CIP priorities of the ligand donor atoms must be determined before the ligands can be placed in the particular positions around the central atom defined by the configuration index.

In the examples presented in Chapter 7, the creation of names and formulae was illustrated with coordination entities that contain only one kind of central atom, principally because there are complications associated with determining central-atom priority for ordering the names and assigning ligands to particular central atoms. Nevertheless, deciphering names with more than one kind of central atom is more straight-forward than producing such names. In these systems the central atoms are all placed at the end of the name. All κ-terms with a 1 placed in front of the κ refer to ligands that are attached to the first central atom listed, while those with 2 refer to the second central atom, and so forth. If there is more than one such number and they are separated by colons, the κ-term must be understood to be referring to a bridging ligand and the numbers identify the central atoms that the ligand bridges. If a formula for a coordination entity is presented with more than one kind of central atom, it may be difficult to identify unequivocally the complex, as the conventions used in formulae are both less specific and less well developed than for those used in names. The process of deciphering names is illustrated below with a number of examples, in order of increasing complexity.

Examples 1–5 below illustrate the methods for name and formula decipherment outlined above.

Examples

1. *cis*-diamminedichloridoplatinum(II) **or** *cis*-[PtCl$_2$(NH$_3$)$_2$]

Identify central atom/ion: platinum(II)

Identify ligands: two ammonia ligands, each coordinating through nitrogen; two chloride ions

Identify geometry: four-coordinate complexes are most commonly either square planar or tetrahedral, but of these the *cis* descriptor can be applied only to the square planar geometry. Platinum(II) complexes are nearly always square-planar.

Draw the complex.

2. (*SPY*-5-12)-dibromidotris[di-*tert*-butyl(phenyl)phosphane-κ*P*]palladium **or** [PdBr$_2$\{P(*t*-Bu)$_2$Ph\}$_3$]

Identify central atom/ion: palladium

Identify ligands: two bromide ions; three di-*tert*-butyl(phenyl)phosphane ligands, represented by P(*t*-Bu)$_2$Ph.

Draw complicated ligand.

Identify geometry: *SPY*-5 indicates square pyramidal geometry with five coordinating atoms; -12 indicates that one of the highest CIP ranked ligands (bromide) is coordinated in the axial site, and that a second ranked ligand is coordinated *trans* to the remaining bromide ligand.

Draw the complex.

3. Pentaamminenitrito-κO-cobalt(III) chloride **or** $[Co(NH_3)_5(ONO)]Cl_2$

Identify central atom/ion: cobalt(III)
Identify ligands: $5 \times$ ammonia ligands; $1 \times$ nitrite ligand, coordinated through oxygen
Identify geometry: six ligands implying octahedral geometry
Two chloride ions required to balance the charge
Draw the complex.

4. (*SPY*-5-21)-tetrakis(μ-acetato-$\kappa O,\kappa O'$)bis[(pyridine)copper(II)] **or**
   $[(py)Cu(\mu-OAc)_4Cu(py)]$

Identify central atoms/ions: $2 \times$ copper(II) ions
Identify ligands: two pyridines, one coordinated to each copper ion; four bridging acetate ligands, with different oxygen atoms coordinated to each copper ion
Identify geometry: *SPY*-5 indicates square-pyramidal geometry with five coordinating atoms; -21 indicates that one of the lower CIP ranked ligands (pyridine) is coordinated in the axial site, the remaining, higher ranked, ligands are attached in the square plane.
Draw the structure.

5. carbonyl-1κC-trichlorido-1$\kappa^2 Cl$,2κCl-bis(triphenylphosphane▶
-1κP)iridiummercury(Ir—Hg) or
[ClHgIr(CO)Cl$_2$(PPh$_3$)$_2$]

Identify central atoms/ions: iridium (central atom 1, because it is first in the list at the end of the name), mercury (central atom 2); these atoms are bonded to each other because the metal-metal bond is indicated at the end of the name, and because they are placed adjacent to each other in the formula.

Identify ligands: one carbonyl ligand, bonded to central atom 1 (iridium), through the carbon atom, as indicated by the κ-term, or placement in the formula; three chloride ligands, two bonded to central atom 1 and one to central atom 2; two triphenylphosphane ligands, both attached to iridium.

Identify geometry: mercury is attached to two atoms, probably linearly; iridium is attached to six atoms, therefore octahedral geometry is likely. There is no information to specify precisely which isomer is present.

Draw the structure.

One other important point to remember is that nomenclature conventions have changed as chemistry evolved. Some features of old names are no longer used in modern nomenclature. For example, it is only recently that ligands such as chloride and oxide have been represented in names of coordination compounds by 'chlorido' and 'oxido'. For many years, those ligands were referred to as 'chloro' and 'oxo'. It is not practical to provide a detailed outline of nomenclature conventions that are no longer in use. Nevertheless even nomenclature systems that have been supplanted are likely to have had logical bases, and it is probable that such bases are similar to those currently recommended. The reader of an old name should be able to make some progress in deciphering it merely by exploiting similarities to currently recommended names.

13.4   DECIPHERING NAMES OF POLYMERIC COMPOUNDS

Polymers or macromolecules have names that are fundamentally different from those of the compounds discussed above. This arises because their structures are normally not clearly defined or even definable, unlike those of the smaller molecules discussed in Sections 13.2 and 13.3. The two principal classes of name, source-based and structure-based, take the form 'poly(monomer)' or 'poly-monomer', and 'poly(constitutional repeating unit)', respectively. Superficially these look similar, but the source unit conveys little or nothing about the structure of the polymer, merely the monomer from which the polymer was obtained, whereas the repeating unit, more properly the constitutional repeating unit (CRU), defines the unit from which the polymer chain is built. Note that although this is indeed structural information it conveys no information about complications such as structural irregularities or chain end-groups. A more detailed discussion about both types of polymer nomenclature is presented in Chapter 9.

In structure-based nomenclature, the CRU is either a divalent organic group (for an organic polymer), or a coordination/organometallic group (for an inorganic polymer). The name of the CRU should therefore be deciphered using the appropriate strategies outlined above for organic or inorganic compounds. The structure is made up of these units linked together, head-to-tail.

Source-based nomenclature provides information on the structure only if the reader realises the structural implications of the formation of the polymer from the cited monomer. Ultimately, a source-based name can only be recognised as such by deciphering the portion of the name within the parentheses and establishing whether it is a compound rather than a divalent fragment.

The Examples in Sections 9.2, 9.3 and 9.4 illustrate the differences between the names that make up sourced-based and structure-based nomenclatures.

13.5 DECIPHERING THE NAMES OF BIOLOGICAL MOLECULES

Biological molecules may be either macromolecules or simpler organic molecules, such as those discussed in Section 13.2, or even something intermediate. However, the disciplines of biology and biochemistry have grown up generally independently of the classical forms of chemistry, and so have the names of the compounds with which they are often concerned. Materials such as fats, lipids, proteins, nucleotides and vitamins have structures that are not conveniently handled by the methods described above, and so independent, though related, classes of nomenclature have developed. The more experienced reader will recognise when such biological materials are being dealt with and will refer to the appropriate authorities when necessary. The present book contains a summary of the various kinds of biological and biochemical name and their application in Chapter 11, where a list of references to relevant authorities is also provided.

14 The IUPAC International Chemical Identifier (InChI)

14.1 IUPAC INTERNATIONAL CHEMICAL IDENTIFIERS

Chemists use diagrammatic representations to convey structural information, and these are sometimes supplemented by verbal descriptions of structure. Conventional chemical nomenclature is a means of specifying a chemical structure in words, and systematic nomenclature provides an unambiguous description of a structure, a diagram of which can be reconstructed from its systematic name. The **IUPAC International Chemical Identifier**, or **InChI**, which is currently being developed, is a machine-readable string of symbols which enables a computer to represent the compound in a completely unequivocal manner. InChIs are produced by computer from structures drawn on-screen, and the original structure can be regenerated from an InChI with appropriate software. An InChI is not directly intelligible to the normal human reader, but InChIs will in time form the basis of an unequivocal and unique database of all chemical compounds.

There is more than one way to specify molecular structures, and those based on 'connection tables' (specifications of atomic connectivities) are more suitable for processing by computer than conventional nomenclature, as they are matrix representations of molecular graphs, readily governed and handled by graph theory. This does not imply that traditional IUPAC nomenclature will eventually be displaced by computer methods and the continued development of verbal nomenclature has run in parallel with the development of InChIs.

The IUPAC International Chemical Identifier (InChI) is a freely available, non-proprietary identifier for chemical substances that can be used in both printed and electronic data sources. It is generated from a computerised representation of a molecular structure diagram produced by chemical structure-drawing software. Its use enables linking of diverse data compilations and unambiguous identification of chemical substances.

A full description of the InChI and of the software for its generation are available from the IUPAC website (reference 1), and further information can be obtained from the website of the InChI Trust (reference 2), a consortium of journal publishers, database providers and chemical software developers constituted in 2009 to provide direction and funding for ongoing development of the InChI standard.

Principles of Chemical Nomenclature: A Guide to IUPAC Recommendations
Edited by G. J. Leigh
© International Union of Pure and Applied Chemistry 2011
Published by the Royal Society of Chemistry, www.rsc.org

A full account of the InChI project is in preparation (reference 3). Commercial structure-drawing software that will generate the Identifier is available from several organisations, which are listed on the IUPAC website (reference 1).

14.2   SHORT DESCRIPTION OF THE InChI

The conversion of structural information to its InChI is based on a set of IUPAC structure conventions and the rules for normalisation and canonicalisation (conversion to a single, predictable sequence) of a structure representation. The resulting InChI is simply a series of characters that serve to identify uniquely the structure from which it was derived. This conversion of a graphical representation of a chemical substance into the unique InChI character string can be carried out automatically by anyone using the freely available programs, and the facility can be built into any program dealing with chemical structures. The InChI uses a layered format to represent all the available structural information relevant to compound identity. The current InChI layers are listed below and are expected to expand as more features are added. Each layer in an InChI representation contains a specific type of structural information. These layers, automatically extracted from the input structure, are designed so that each successive layer adds additional detail to the Identifier. The specific layers generated depend on the level of structural detail available and whether or not allowance is made for tautomerism. Of course, if there are any ambiguities or uncertainties in the original structure representation, these will remain in the InChI.

This layered structure design of an InChI offers a number of advantages. If two structures for the same substance are drawn at different levels of detail, the one with the lower level of detail will, in effect, be contained within the other. Specifically, if one substance is drawn with stereo-bonds and the other without, the layers in the latter will be a subset of the former. The same will hold for compounds treated by one author as tautomers and by another as exact structures with all hydrogen atoms fixed. This can work at a finer level. For example, if one author includes a double bond and tetrahedral stereochemistry, but another omits any reference to stereochemistry, the InChI for the latter description will be contained within that for the former.

14.3   THE STRUCTURE OF InChIs

The successive layers of an InChI are characterised as follows.
1. Formula
2. Connectivity (no formal bond orders)
   a. disconnected metals
   b. connected metals
3. Isotopes
4. Stereochemistry
   a. double bond
   b. tetrahedral
5. Tautomers (on or off)

Note that charges are not considered within the basic InChI, but are added at the end of the InChI string, as in Example 2 below.

Two examples of InChI representations are given below. However, it is important to recognise that InChI strings are intended for use by computers and end-users need not understand any of their details. In fact, the open nature of InChI and its flexibility of representation, after implementation into software systems, may allow chemists to be even less concerned with the details of structure representation by computers.

Examples

1. Name: guanine

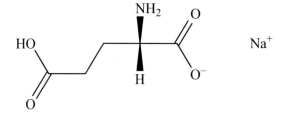

The InChI for this structure is:
InChI=1/C5H5N5O/c6-5-9-3-2(4(11)10-5)7-1-8-3/h1H,(H4,6,7,8,9,10,11)/f/h8,10H,6H2

2. Name: sodium L-glutamate(1−)

The InChI for this structure is:
InChI=1/C5H9NO4.Na/c6-3(5(9)10)1-2-4(7)8;/h3H,1-▶
2,6H2,(H,7,8)(H,9,10);/q;+1/p-1/t3-;/m1./s1/fC5H8NO4.Na/h7H;/q-1;m

The layers in the InChI string are separated by the slash, /, followed by a lower-case letter (except for the first layer, the chemical formula), with the layers arranged in a predefined order. In the Examples above the following segments are included:

InChI version number
/chemical formula
/c connectivity-1.1 (excluding terminal H)
/h connectivity-1.2 (locations of terminal H, including mobile H attachment points)
/q charge

/p proton balance
/t tetrahedral parity
/m parity inverted to obtain relative stereo (1 = inverted, 0 = not inverted)
/s stereo type (1 = absolute, 2 = relative, 3 = racemic)
/f chemical formula of the fixed-H structure if it is different
/h connectivity-2 (locations of fixed mobile H)

One of the most important applications of InChI is the facility to locate literature mention of a chemical substance using internet-based search engines. This is made easier by using a shorter (compressed) form of InChI, known as an InChIKey. The InChIKey is a 27-character representation that, because it is compressed, cannot be reconverted into the original structure, but it is not subject to the undesirable and unpredictable breaking of longer character strings by all current search engines. The usefulness of the InChIKey as a search tool is enhanced if it is derived from a 'standard' InChI, *i.e.,* an InChI produced with standard option settings for features such as tautomerism and stereochemistry.

An Example is shown below; the standard InChI is denoted by the letter S after the version number. Use of the InChIKey also allows searches based solely on atomic connectivity (first 14 characters). The software for generating InChIKey is also available from the IUPAC website (reference 1).

Example

3. Name: caffeine

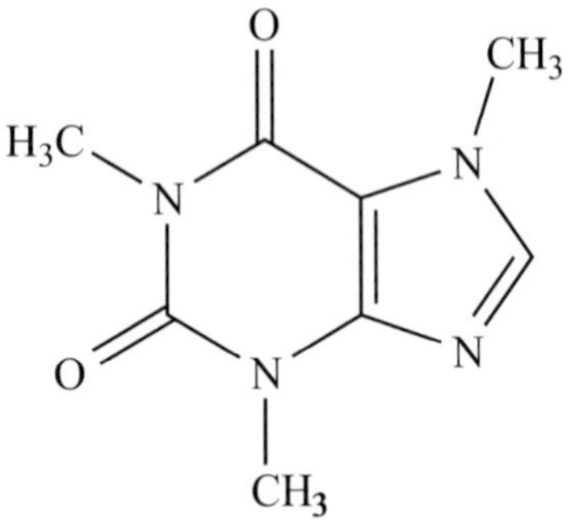

InChI=1S/C8H10N4O2/c1-10-4-9-6-5(10)7(13)12(3)8(14)11(6)2/h4H,1-3H3

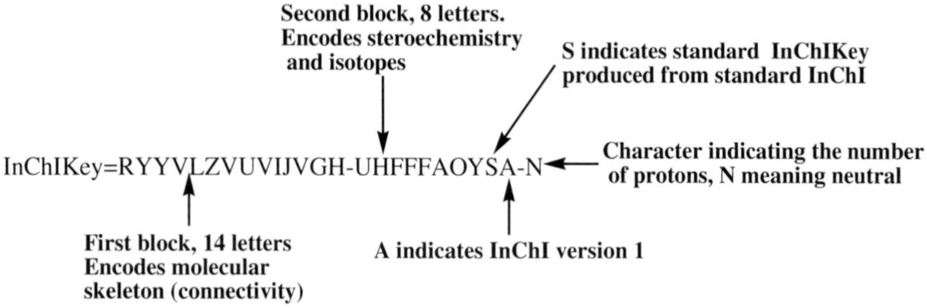

Second block, 8 letters.
Encodes steroechemistry
and isotopes

S indicates standard InChIKey
produced from standard InChI

InChIKey=RYYVLZVUVIJVGH-UHFFFAOYSA-N

Character indicating the number
of protons, N meaning neutral

First block, 14 letters
Encodes molecular
skeleton (connectivity)

A indicates InChI version 1

The enormous databases compiled by organisations such as PubChem (reference 4), the US National Cancer Institute, and ChemSpider (reference 5) contain millions of InChIs and InChIKeys, which allow sophisticated searching of these collections. PubChem provides InChI-based structure-search facilities (for both identical and similar structures) (reference 6), and ChemSpider offers both search facilities and web services enabling a variety of InChI and InChIKey conversions (reference 7). The NCI Chemical Structure Lookup Service (reference 8) provides InChI-based search access to over 74 million chemical structures from over 100 different public and commercial data sources. Members of the InChI Trust (May, 2011) include Accelrys, Dialog, Elsevier, FIZ-Chemie Berlin, IUPAC, Nature, Royal Society of Chemistry, Springer, Wiley, Taylor and Francis, IBM Research, ACD Labs, Chemaxon, and OpenEye, and this list is expected to grow.

In the age of the computer, the IUPAC International Chemical Identifier is an essential component of the chemist's armoury of information tools, allowing location and manipulation of chemical data with unprecedented ease and precision.

REFERENCES

1. http://www.iupac.org/inchi
2. http://www.inchi-trust.org
3. *Pure and Applied Chemistry*, in preparation.
4. http://pubchem.ncbi.nlm.nih.gov
5. http://www.chemspider.com
6. http://pubchem.ncbi.nlm.nih.gov/search
7. http://www.chemspider.com/InChI.asmx
8. http://cactus.nci.nih.gov/cgi-bin/lookup/search

15 Nomenclature in the Making

Nomenclature is not a static subject. It changes as new kinds of compound are synthesised and new procedures and devices have to be invented. Fullerenes are a recent case in point, and their detailed nomenclature has only recently been codified (*Pure and Applied Chemistry,* 2002, **74**(4), 629–695; *Pure and Applied Chemistry,* 2005, **77**(5), 843–923). Even established procedures are being reviewed continuously by IUPAC and revised, but the principles described in the current edition of *Principles* are not likely to be amended to any significant degree.

Precise application of nomenclature rules may not be easy for beginners, especially for naming complex structures. That is why IUPAC has published a series of guides and compendia to help and guide both new and experienced chemists. The current version of the *Nomenclature of Organic Chemistry* (the Blue Book) which was published in 1979 is a somewhat daunting compendium. A new edition is currently (2011) in a late stage of preparation. However, the 1993 edition of *A Guide to IUPAC Nomenclature of Organic Compounds* is unlikely to be rendered significantly redundant by the changes announced since 1979. Most of it will be incorporated in the new edition of the Blue Book.

Inorganic nomenclature has traditionally been less well codified than organic nomenclature, for a variety of reasons. First, attempts at general codification began much later than those dealing with organic nomenclature. The first IUPAC rules for inorganic nomenclature were published as late as 1940. In contrast, the Geneva meeting that started the process with organic nomenclature took place in 1892. Inorganic nomenclature also deals with aspects of chemistry such as oxidation states and configurations of inorganic structures which were not important or not even recognised by the fathers of organic chemistry. The most recent edition of the *IUPAC Nomenclature of Inorganic Chemistry* (the Red Book) was published in 2005, and it incorporates many recommendations that are new when compared to the earlier 1990 edition. Some of these, such as the change of the names of ligands such as Cl from 'chloro' to 'chlorido' may be thought unnecessary by some chemists. However, the intention is to make the treatment of such species standard throughout coordination nomenclature. In any case, the older names will persist in the literature for a long time to come, so it is currently acceptable to use either an 'yl' or an 'ido' suffix when naming a ligand such as CH_3. The recommendation that the compound PH_3 should now be called 'phosphane' rather than the traditional 'phosphine' is to standardise the application of organic substitutive nomenclature to parent hydrides across both inorganic and organic chemistry.

Principles of Chemical Nomenclature: A Guide to IUPAC Recommendations
Edited by G. J. Leigh
© International Union of Pure and Applied Chemistry 2011
Published by the Royal Society of Chemistry, www.rsc.org

Since the previous edition of *Principles* was published in 1998, IUPAC has undergone a considerable reorganisation. In particular, many of the old Commissions, and particularly those dealing with organic, inorganic and polymer nomenclatures, have been abolished. Whereas they were all committees of separate IUPAC Divisions, their functions are now the remit of a single Division, the Chemical Nomenclature and Structure Representation Division, in IUPAC jargon Division VIII. This has enabled the IUPAC experts to take a much more unified view of nomenclature and this has been reflected in the content of the current edition of *Principles*, and with the inclusion of enlarged contributions on topics such as organometallic and biochemical nomenclatures. In addition, the treatment is much more unified than in the first edition.

IUPAC is also working on other developments that will be of value not only to the chemistry community but to the chemistry and commercial world at large. These include the IUPAC International Chemical Identifier (InChI), which will make the searching of a variety of independent databases much easier and more fruitful. The project to produce Preferred IUPAC Names (PINs) is already well advanced for organic names and work has been started on inorganic names. The ultimate aim is to reduce the multiplicity of names which are sometimes found for any given compound to a single example, and this name should reflect the standard nomenclature methods applied in the English language across chemistry.

16 Guidelines for Drawing Chemical Structures

In the presentation of the material in this book, care has been taken to show diagrams and chemical structures that are clear and easily comprehensible. Drawing of chemical structures in a way that enables rapid and easy comprehension by a reader can be a complex and difficult process, especially if detailed organic or inorganic stereochemistry is to be represented. IUPAC has published a series of recommendations on how best to do this, namely Graphical Representation Standards for Chemical Structure Diagrams (IUPAC Recommendations 2008), *Pure and Applied Chemistry*, 2008, **80**(2), 277–410, and Graphical Representation of Stereochemical Configuration (IUPAC Recommendations 2006), *Pure and Applied Chemistry*, 2006, **78**(10), 1897–1970. The interested reader is referred to these publications for detailed information, but it will be helpful to anyone setting out to draw chemical structures to bear certain considerations in mind.

The two extremely useful references cited above make the following suggestions. "The purpose of a chemical structure diagram is to convey information – typically the identity of a molecule – to another human reader or as input to a computer program ... Know your audience ... Avoid ambiguous drawing styles ... Avoid inconsistent drawing styles."

Two important requirements should be remembered when drawing chemical structures: structure diagrams must be drawn correctly, and they should be drawn well. This is similar to the requirement of using the correct word in spoken or written language. At its most obvious, it would not be acceptable to use a diagram of aniline, Example 1, to illustrate a discussion of phenol, Example 2.

Examples

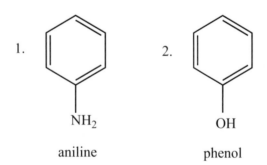

1.

2.

aniline

phenol

Principles of Chemical Nomenclature: A Guide to IUPAC Recommendations
Edited by G. J. Leigh
© International Union of Pure and Applied Chemistry 2011
Published by the Royal Society of Chemistry, www.rsc.org

The diagram of phenol can be drawn in many other ways, all equally correct. For example, the diagram can be rotated in the plane of the paper or scaled to a different size. The ring could be distorted so that it is a non-regular hexagon, the individual bonds could be displayed with different thicknesses, and the OH atom label could be drawn in bold, or italic, or in a font other than that used here. Each of those diagrams might be an acceptable representation of phenol.

Examples

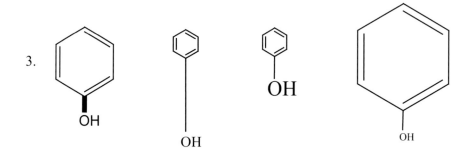

Drawing structures well is also important for another reason. Examples 3 are all unacceptable depictions of the phenol molecule, for reasons that should be self-evident. One might argue that each of them is "correct", but each of them is also an unusual representation. Rather than helping convey information, structure diagrams drawn in unusual ways are often confusing, simply because they force the reader to interpret something not often seen. Structure diagrams written in unusual ways, even when the diagrams are otherwise accurate, are not acceptable.

The guidelines to be followed in order to produce acceptable structure diagrams are as follows.

• Draw clearly, for example, by using consistent bond lengths and angles, and common fonts such as Times New Roman used in this book.

• The selected font size should make comfortable reading.

• Line widths should be consistent throughout the diagram.

• Bonds should be clearly designated and lengths kept standard.

• Atom labels should use conventional presentation and consistent sizes. Multi-atom labels, such as COOH, should be on a single line and the last cited atom closest to the point at which the label joins the rest of the formula, as with –COOH or HOOC–, as appropriate.

• Abbreviations for groups of atoms and in labels should be unambiguous and readily understood. It may be necessary to define them somewhere in the text.

• Bond angles should be represented as consistently as possible.

• Formalisms to represent structures in three dimensions should be self-consistent and in accordance with accepted practice.

Appendix: Tables of General Application to the Whole Text

Table P1 The elements and their symbols, as approved by IUPAC in 2010.
The names and symbols of the elements with atomic numbers of 113 or greater are temporary.

Atomic Number	Symbol	Name	Atomic Number	Symbol	Name
1	H	hydrogen[a]	44	Ru	ruthenium
2	He	helium	45	Rh	rhodium
3	Li	lithium	46	Pd	palladium
4	Be	beryllium	47	Ag	silver[d]
5	B	boron	48	Cd	cadmium
6	C	carbon	49	In	indium
7	N	nitrogen[b]	50	Sn	tin[d]
8	O	oxygen	51	Sb	antimony[d]
9	F	fluorine	52	Te	tellurium
10	Ne	neon	53	I	iodine
11	Na	sodium[d]	54	Xe	xenon
12	Mg	magnesium	55	Cs	caesium[c]
13	Al	aluminium[c]	56	Ba	barium
14	Si	silicon	57	La	lanthanum
15	P	phosphorus	58	Ce	cerium
16	S	sulfur[b,c]	59	Pr	praseodymium
17	Cl	chlorine	60	Nd	neodymium
18	Ar	argon	61	Pm	promethium
19	K	potassium[d]	62	Sm	samarium
20	Ca	calcium	63	Eu	europium
21	Sc	scandium	64	Gd	gadolinium
22	Ti	titanium	65	Tb	terbium
23	V	vanadium	66	Dy	dysprosium
24	Cr	chromium	67	Ho	holmium
25	Mn	manganese	68	Er	erbium
26	Fe	iron[d]	69	Tm	thulium
27	Co	cobalt	70	Yb	ytterbium
28	Ni	nickel	71	Lu	lutetium
29	Cu	copper[d]	72	Hf	hafnium
30	Zn	zinc	73	Ta	tantalum
31	Ga	gallium	74	W	tungsten[e]
32	Ge	germanium	75	Re	rhenium
33	As	arsenic	76	Os	osmium
34	Se	selenium	77	Ir	iridium
35	Br	bromine	78	Pt	platinum
36	Kr	krypton	79	Au	gold[d]
37	Rb	rubidium	80	Hg	mercury[d]
38	Sr	strontium	81	Tl	thallium
39	Y	yttrium	82	Pb	lead[d]
40	Zr	zirconium	83	Bi	bismuth
41	Nb	niobium	84	Po	polonium
42	Mo	molybdenum	85	At	astatine
43	Tc	technetium	86	Rn	radon

Principles of Chemical Nomenclature: A Guide to IUPAC Recommendations
Edited by G. J. Leigh
© International Union of Pure and Applied Chemistry 2011
Published by the Royal Society of Chemistry, www.rsc.org

APPENDIX

Table P1 *(Continued)*

Atomic Number	Symbol	Name	Atomic Number	Symbol	Name
87	Fr	francium	104	Rf	rutherfordium
88	Ra	radium	105	Db	dubnium
89	Ac	actinium	106	Sg	seaborgium
90	Th	thorium	107	Bh	bohrium
91	Pa	protactinium	108	Hs	hassium
92	U	uranium	109	Mt	meitnerium
93	Np	neptunium	110	Ds	darmstadtium
94	Pu	plutonium	111	Rg	roentgenium
95	Am	americium	112	Cn	copernicium
96	Cm	curium	113	Uut	ununtrium[f]
97	Bk	berkelium	114	Uuq	ununquadium[f]
98	Cf	californium	115	Uup	ununpentium[f]
99	Es	einsteinium	116	Uuh	ununhexium[f]
100	Fm	fermium	117	Uus	ununseptium[f]
101	Md	mendelevium	118	Uuo	ununoctium[f]
102	Nb	nobelium	119	Uue	ununennium[f]
103	Lr	lawrencium	120	Ubn	unbinilium[f]

[a] The hydrogen isotopes ^2H and ^3H are named deuterium and tritium, respectively, for which the symbols D and T may be used, though not generally recommended because they may require citing the symbols in different places when alphabetic ordering is used.

[b] The name 'azote', used in French for nitrogen, provides the root for 'az' used in naming some nitrogen compounds. The Greek-based name 'theion' provides the root for 'thi' used in naming some sulfur compounds.

[c] The alternative spelling 'aluminum' and related pronunciation are used in the United States of America, though not recommended by IUPAC. The name 'caesium' is sometimes spelled 'cesium', and 'sulfur' spelled 'sulphur', though neither alternative is recommended by IUPAC.

[d] For reasons of history and usage, the element symbols, Na, K, Fe, Cu, Ag, Sn, Sb, Au, Hg, and Pb are derived from the Latinate names natrium, kalium, ferrum, cuprum, argentum, stannum, stibium, aurum, hydrargyrum, and plumbum, respectively.

[e] For historical reasons, the element symbol W is derived from the name of the ore, wolframite, from which the metal tungsten was originally isolated. The name wolfram is still used for tungsten in some languages, but is no longer recommended for use in English.

[f] These names are based upon the corresponding atomic numbers. Claims have been made to have synthesised some of those listed here, for example Uuh and Uuo. If and when such claims have been assessed by the relevant IUPAC/IUPAP processes and are accepted, the synthesisers will then have the right to suggest a new, permanent name. Until then all these names and three-letter symbols are available for use, but are to be regarded as temporary, and will be revised when any claim to have prepared such elements is accepted. See *Pure and Applied Chemistry*, 1979, **51**(2), 381–384 and *Pure and Applied Chemistry*, 2002, **74**(5), 787–791.

Table P2 A Periodic Table, adapted from the *Nomenclature of Inorganic Chemistry*, Recommendations 2005.

1																	18
1 **H**	2												13	14	15	16	17	2 **He**
3 **Li**	4 **Be**												5 **B**	6 **C**	7 **N**	8 **O**	9 **F**	10 **Ne**
11 **Na**	12 **Mg**	3	4	5	6	7	8	9	10	11	12		13 **Al**	14 **Si**	15 **P**	16 **S**	17 **Cl**	18 **Ar**
19 **K**	20 **Ca**	21 **Sc**	22 **Ti**	23 **V**	24 **Cr**	25 **Mn**	26 **Fe**	27 **Co**	28 **Ni**	29 **Cu**	30 **Zn**	31 **Ga**	32 **Ge**	33 **As**	34 **Se**	35 **Br**	36 **Kr**
37 **Rb**	38 **Sr**	39 **Y**	40 **Zr**	41 **Nb**	42 **Mo**	43 **Tc**	44 **Ru**	45 **Rh**	46 **Pd**	47 **Ag**	48 **Cd**	49 **In**	50 **Sn**	51 **Sb**	52 **Te**	53 **I**	54 **Xe**
55 **Cs**	56 **Ba**	*57-71 lanthanoids	72 **Hf**	73 **Ta**	74 **W**	75 **Re**	76 **Os**	77 **Ir**	78 **Pt**	79 **Au**	80 **Hg**	81 **Tl**	82 **Pb**	83 **Bi**	84 **Po**	85 **At**	86 **Rn**
87 **Fr**	88 **Ra**	‡89-103 actinoids	104 **Rf**	105 **Db**	106 **Sg**	107 **Bh**	108 **Hs**	109 **Mt**	110 **Ds**	111 **Rg**	112 **Cn**	113 Uut	114 Uuq	115 Uup	116 Uuh	117 Uus	118 Uuo

*57 **La**	58 **Ce**	59 **Pr**	60 **Nd**	61 **Pm**	62 **Sm**	63 **Eu**	64 **Gd**	65 **Tb**	66 **Dy**	67 **Ho**	68 **Er**	69 **Tm**	70 **Yb**	71 **Lu**
‡89 **Ac**	90 **Th**	91 **Pa**	92 **U**	93 **Np**	94 **Pu**	95 **Am**	96 **Cm**	97 **Bk**	98 **Cf**	99 **Es**	100 **Fm**	101 **Md**	102 **No**	103 **Lr**

Only atomic numbers are included here in addition to atomic symbols. Atomic weights are under continuous review by IUPAC and are subject to regular revision. For the latest values, readers are referred to *Pure and Applied Chemistry*, 2011, **83**(2), 359–396 and subsequent IUPAC publications. Note that IUPAC recommends Group numbering which is consistent with this form of the Table.

APPENDIX

Table P3 The element sequence used to determine relative electronegativities **for nomenclature purposes**. This sequence should not be used for assessing electronegativities for any other purpose, since its only objective is to ensure that element sequences are quoted consistently when constructing chemical names. The sequence starts at top right (most electronegative) and finishes at bottom left (least electronegative, or most electropositive).

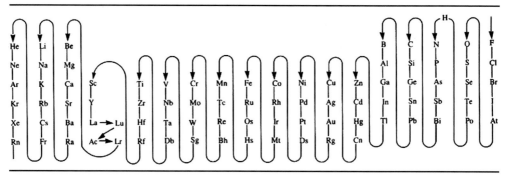

Table P4 Some common shapes encountered in organic and inorganic structures. The left hand column shows the 'polyhedra' with groups attached at each apex. The right hand column shows the equivalent representation with bonds represented in three dimensions, as described in the text. The spatial orientations in the two columns are not in all cases exactly equivalent.

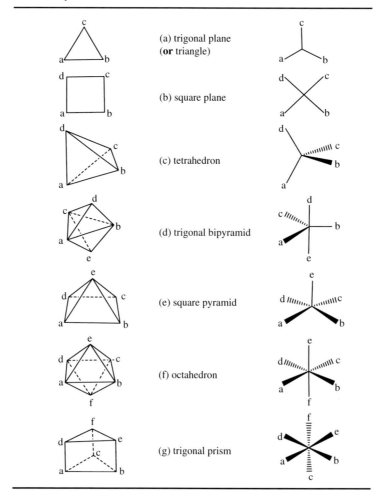

(a) trigonal plane (**or** triangle)

(b) square plane

(c) tetrahedron

(d) trigonal bipyramid

(e) square pyramid

(f) octahedron

(g) trigonal prism

Table P5 The shapes and polyhedral symbols of some commonly encountered coordination polyhedra.

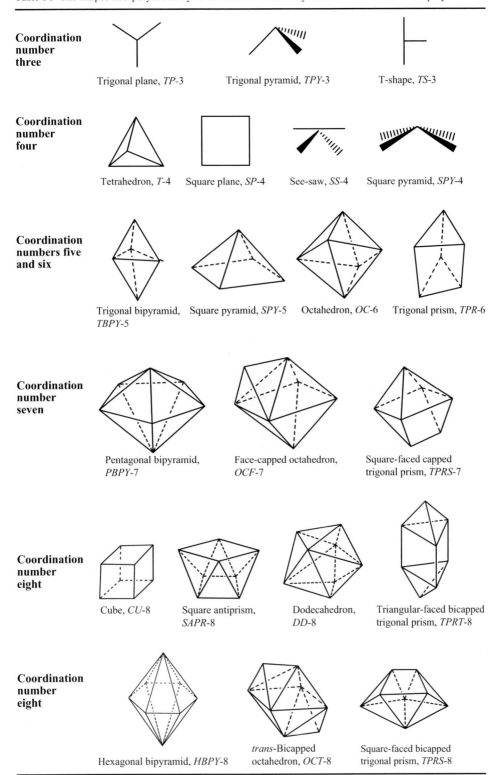

Coordination number three		
Trigonal plane, *TP*-3	Trigonal pyramid, *TPY*-3	T-shape, *TS*-3

Coordination number four			
Tetrahedron, *T*-4	Square plane, *SP*-4	See-saw, *SS*-4	Square pyramid, *SPY*-4

Coordination numbers five and six

Trigonal bipyramid, *TBPY*-5 — Square pyramid, *SPY*-5 — Octahedron, *OC*-6 — Trigonal prism, *TPR*-6

Coordination number seven

Pentagonal bipyramid, *PBPY*-7 — Face-capped octahedron, *OCF*-7 — Square-faced capped trigonal prism, *TPRS*-7

Coordination number eight

Cube, *CU*-8 — Square antiprism, *SAPR*-8 — Dodecahedron, *DD*-8 — Triangular-faced bicapped trigonal prism, *TPRT*-8

Coordination number eight

Hexagonal bipyramid, *HBPY*-8 — *trans*-Bicapped octahedron, *OCT*-8 — Square-faced bicapped trigonal prism, *TPRS*-8

For further explanation, see Section 7.5.7. This table is not exhaustive, nor should it be taken to imply that in the real world all coordination polyhedra are regular and undistorted. In the diagrams, the dotted lines indicate polyhedral edges that are present, but would not be visible without rotating the polyhedron or changing the position of view.

APPENDIX

Table P6 Some numerical or multiplicative prefixes encountered in nomenclature.

These are placed before a part of a name to indicate the presence of more than one chemical moiety of a given kind in the substance being considered. All the simple prefixes for 4 or greater have 'kis' variants, but only a selection has been included here.

Number	Numerical prefix	Number	Numerical prefix
1	mono	21	henicosa
2	di (**or** bis)	22	docosa
3	tri (**or** tris)	23	tricosa
4	tetra (**or** tetrakis)	30	triaconta
5	penta (**or** pentakis)	31	hentriaconta
6	hexa (**or** hexakis)	34	tetratriaconta
7	hepta (**or** heptakis)	35	pentatriaconta
8	octa (**or** octakis)	40	tetraconta
9	nona (**or** nonakis)	46	hexatetraconta
10	deca (**or** decakis)	47	heptatetraconta
11	undeca	50	pentaconta
12	dodeca	58	octapentaconta
13	trideca	59	nonapentaconta
14	tetradeca	60	hexaconta
15	pentadeca	70	heptaconta
16	hexadeca	80	octaconta
17	heptadeca	90	nonaconta
18	octadeca	100	hecta
19	nonadeca		
20	icosa		

Table P7 Selected names of homoatomic, binary and certain other simple molecules, ions, compounds, radicals and substituent groups.

This Table may be used as a reference for names of simple compounds and as a source of examples to guide in the naming of further compounds, but it is not comprehensive. The listing is alphabetic in order of the formulae given in the column at the far left. Formulae for **binary** species are written there in an order consistent with the relative positions of the two elements in Table P3, so that ammonia, for example, is found under NH_3, but selane under H_2Se. The only exceptions made to this are the binary species of oxygen and either chlorine or iodine. The formulae of **ternary** and **quaternary** compounds are written strictly alphabetically, e.g., HOCN is found under CHNO and HPO_4 under HO_4P. In the columns to the right of the left-hand column, special formats may be used for formulae in order to stress a particular structure, e.g., under the entry $BrHO_3$ one finds $HOBrO_2$ rather than $HBrO_3$ or $[BrO_2(OH)]$. The symbol ▶ is used for dividing names when this is made necessary by a line break. When the name is reconstructed from the name given in the Table, this symbol should be omitted and the parts joined with no space. All **hyphens** in the Table are true parts of the names. The symbols > and < placed next to an element symbol each denote two single bonds connecting the atom in question to two other atoms and = similarly placed represents a double bond.

For a given compound, various systematic names are supplied and these may be stoichiometric, substitutive, or additive names (see Chapter 4). The reader should be able to recognise the various kinds of name. Some names which are not entirely systematic (or not formed according to the systems mentioned above) but which are still considered acceptable are also supplied. No order of preference is implied by the order in which formulae and names are listed. Where more than one name is supplied, the reader must select that which is most appropriate to the usage and context in which it will be employed. For example, a stoichiometric name may be adequate to specify a compound for which structural details are not required, whereas a parent-hydride-based name or an additive name may be more appropriate in other circumstances.

It may be necessary to browse the Table to find (families of) compounds that match those of interest. Few simple carbon-containing compounds and substituent groups are included. In particular, organic ligands belonging to the general classes alcoholates, thiolates, phenolates, carboxylates, amines, phosphanes and arsanes as well as (partially) dehydronated amines, phosphanes and arsanes are generally *not* included. Their naming is described and exemplified in Chapter 6.

In most formulae we have given complete designations for charge, unpaired electrons, and bonds between elements of the same kind. In daily commerce it may not be necessary to use all these all the time, because such information may be understood or superfluous. For example, indication of the N-N bond in hydrazine may be necessary to convey a complete structural formula of the molecule, but most chemists will be aware of its presence and will understand that it is present without the necessity to depict it. Teachers and writers should use their judgement and supply as much information as the students or recipients need, rather than stick slavishly to the complete unequivocal descriptions.

Note that throughout this Table formulae are presented with the right superscript immediately over the right subscript, as in AsO_4^{3-} and CH_3^-. As discussed in Chapter 2, page 8, the presentations $AsO_4{}^{3-}$ and $CH_3{}^-$ in these and all similar cases, with the right superscript offset, are often clearer and therefore may be preferred.

Formula for uncharged atom or group	Name			
	Uncharged atoms or molecules (including zwitterions and radicals) or substituent groups[a]	Cations (including cation radicals) or cationic substituent groups[a]	Anions (including anion radicals) or anionic substituent groups[b]	Ligands[c]
Ag	silver	silver	argentide	argentido
Al	aluminium	aluminium (general) Al^+ aluminium(1+) Al^{3+} aluminium(3+)	aluminide (general) Al^- aluminide(1−)	aluminido (general) Al^- aluminido(1−)
$AlCl_3$ (see also Al_2Cl_6)	aluminium trichloride **or** trichloroalumane **or** trichloridoaluminium			
AlH_3	aluminium trihydride **or** alumane (parent hydride name) **or** trihydridoaluminium	$AlH_3^{•+}$ alumaniumyl **or** trihydridoaluminium(•1+)	$AlH_3^{•-}$ alumanuidyl **or** trihydridoaluminate(•1−)	
Al_2Cl_6	$[Cl_2Al(\mu\text{-}Cl)_2AlCl_2]$ di-μ-chloridobis▶ (dichloridoaluminium)			

Table P7 (*Continued*)

Formula for uncharged atom or group	Name			
	Uncharged atoms or molecules (including zwitterions and radicals) or substituent groups[a]	Cations (including cation radicals) or cationic substituent groups[a]	Anions (including anion radicals) or anionic substituent groups[b]	Ligands[c]
As	arsenic >As– arsanetriyl	arsenic	arsenide (general) As^{3-} arsenide(3–) **or** arsanetriide	arsenido (general) As^{3-} arsenido(3–) **or** arsanetriido
AsH_3	arsenic trihydride **or** arsane (parent hydride name) **or** trihydridoarsenic	$AsH_3^{\bullet+}$ arsaniumyl **or** trihydridoarsenic($\bullet$1+) –AsH_3^+ arsaniumyl	$AsH_3^{\bullet-}$ arsanuidyl **or** trihydridoarsenate($\bullet$1–)	$AsH_3^{\bullet-}$ trihydridoborato($\bullet$1–)
AsH_4	–AsH_4 λ^5-arsanyl	AsH_4^+ arsanium **or** tetrahydridoarsenic(1+)		
AsH_5	arsenic pentahydride **or** λ^5-arsane (parent hydride name) **or** pentahydridoarsenic			
AsO_3			AsO_3^{3-} trioxidoarsenate(3–) **or** arsenite **or** arsorite –As(=O)(O)$_2$ dioxidooxo-λ^5-arsanyl **or** arsonato	AsO_3^{3-} trioxidoarsenato(3–) **or** arsenito **or** arsorito
AsO_4			AsO_4^{3-} tetraoxidoarsenate(3–) **or** arsenate **or** arsorate	AsO_4^{3-} tetraoxidoarsenato(3–) **or** arsenato **or** arsorato
AsS_4			AsS_4^{3-} tetrasulfidoarsenate(3–)	AsS_4^{3-} tetrasulfidoarsenato(3–)
Au	gold	gold (general) **or** Au^+ gold(1+) Au^{3+} gold(3+)	auride	aurido
B	boron >B– boranetriyl	boron (general) B^+ boron(1+) B^{3+} boron(3+)	boride (general) B^- boride(1–) B^{3-} boride(3–)	borido (general) B^- borido(1–) B^{3-} borido(3–)
BH_2	–BH_2 boranyl	BH_2^+ boranylium **or** dihydridoboron(1+)	BH_2^- boranide **or** dihydridoborate(2–)	BH_2^- boranido **or** dihydridoborato(2–)
BH_3	boron trihydride **or** borane (parent hydride name) **or** trihydridoboron	$BH_3^{\bullet+}$ boraniumyl **or** trihydridoboron($\bullet$1+)	$BH_3^{\bullet-}$ boranuidyl **or** trihydridoborate($\bullet$1–) –BH_3 boranuidyl	$BH_3^{\bullet-}$ trihydridoborato($\bullet$1–)
BH_3O_3	$H_3BO_3 = B(OH)_3$ trihydroxidoboron **or** boric acid			

		BH_4^+ boranium or tetrahydridoboron(1+)	BH_4^- boranuide or tetrahydridoborate(1−)	BH_4^- boranuido or tetrahydridoborato (1−)
BH_4				
BO_3			BO_3^{3-} trioxidoborate(3−) or borate	BO_3^{3-} trioxidoborato(3−) or borato
Ba	barium		baride	barido
Be	beryllium (general) Be^+ beryllium(1+) Be^{2+} beryllium(2+)		beryllide	beryllido
Bi	bismuth		bismuthide (general) Bi^{3-} bismuthide(3−) or bismuthanetriide or bismuthide	bismuthido (general) Bi^{3-} bismuthido(3−) or bismuthanetriido or bismuthido
BiH_3	BiH_3 bismuth trihydride or bismuthane (parent hydride name) or trihydridobismuth $=BiH_3$ λ^5-bismuthanylidene	$BiH_3^{\bullet+}$ bismuthaniumyl or trihydridobismuth(•1+)	$BiH_3^{\bullet-}$ bismuthanuidyl or trihydridobismuthate(•1−)	
BiH_4		BiH_4^+ bismuthanium or tetrahydridobismuth(1+)		
Bi_5		Bi_5^{4+} pentabismuth(4+)		
Br	bromine (general) $Br^\bullet$ bromine(•) or monobromine –Br bromo	bromine (general) Br^+ bromine(1+)	bromide (general) Br^- bromide(1−) or bromide	bromido (general) Br^- bromido(1−) or bromido
BrCN	BrCN cyanobromane or bromidonitridocarbon			
BrH, see HBr				
$BrHO_3$	$HOBrO_2$ hydroxy-λ^5-bromanedione or hydroxidodioxidobromine or bromic acid			
$BrHO_4$	$HOBrO_3$ hydroxy-λ^7-bromanetrione or hydroxidotrioxidobromine or perbromic acid			
BrH_2, see H_2Br				
Br_2	Br_2 dibromine	$Br_2^{\bullet+}$ dibromine(•1+)	$Br_2^{\bullet-}$ dibromide(•1−)	Br_2 dibromine $Br_2^{\bullet-}$ dibromido(•1−)
C	carbon (general) C monocarbon >C< methanetetrayl =C= methanediylidene	carbon (general) C^+ carbon(1+)	carbide (general) C^- carbide(1−) C^{4-} carbide(4−) or methanetetraide	carbido (general) C^- carbido(1−) C^{4-} methanetetrayl or methanetetraido or carbido(4−)

Table P7 (*Continued*)

Formula for uncharged atom or group	Name			
	Uncharged atoms or molecules (including zwitterions and radicals) or substituent groups[a]	Cations (including cation radicals) or cationic substituent groups[a]	Anions (including anion radicals) or anionic substituent groups[b]	Ligands[c]
CCl_2O	$COCl_2$ carbon dichloride oxide **or** dichloridooxidocarbon **or** carbonyl dichloride			
CHN	HCN hydrogen cyanide **or** methanenitrile **or** hydridonitridocarbon **or** formonitrile; HNC carbidohydridonitrogen; $>C=NH$ carbonimidoyl; $=C=NH$ iminomethylidene			
$CHNO$	$HCNO$ formonitrile oxide **or** (hydridocarbonato)oxidonitrogen; $HOCN$ cyanic acid **or** hydroxidonitridocarbon; $HNCO$ isocyanic acid **or** (hydridonitrato)▶ oxidocarbon		$HNCO^{\bullet-}$ (hydridonitrato)oxidocarbonate(•1−); $HOCN^{\bullet-}$ hydroxidonitridocarbonate(•1−)	$HNCO^{\bullet-}$ (hydridonitrato)oxidocarbonato(•1−); $HOCN^{\bullet-}$ hydroxidonitridocarbonato(•1−)
CH_2	CH_2 λ^2-methane; $CH_2^{\bullet}$ methylidene **or** carbene; $>CH_2$ methylene; $=CH_2$ methylidene		CH_2^{2-} methanediide **or** dihydridocarbonate(2−); $-CH_2$ methanidyl	CH_2^{2-} methanediido **or** dihydridocarbonato(2−)
CH_3	$CH_3^{\bullet}$ methyl; $-CH_3$ methyl	CH_3^+ methylium **or** trihydridocarbon(1+)	CH_3^- methanide **or** trihydridocarbonate(1−)	CH_3 methyl **or** methanido **or** trihydridocarbonato(1−)
CH_4	CH_4 tetrahydridocarbon **or** methane (parent hydride name)	$CH_4^{+\bullet}$ methaniumyl **or** tetrahydridocarbon(•1+)	$CH_4^{-\bullet}$ methanuidyl **or** tetrahydridocarbonate(•1−)	
CH_5		CH_5^+ methanium **or** pentahydridocarbon(1+)		
CN	$CN^{\bullet}$ nitridocarbon(•); $-CN$ cyano; $-NC$ isocyano	CN^+ nitridocarbon(1+)	CN^- cyanide **or** nitridocarbonate(1−)	nitridocarbonato(1−)-κC **or** cyanido
CO	CO carbon mon(o)oxide; $>C=O$ carbonyl	$CO^{+\bullet}$ oxidocarbon(•1+); CO^{2+} oxidocarbon(2+)	$CO^{-\bullet}$ oxidocarbonate(•1−)	CO carbonyl
CNS	$SCN^{\bullet}$ nitridosulfidocarbon(•); $-SCN$ thiocyanato; $-NCS$ isothiocyanato		SCN^- nitridosulfidocarbonate(1−) **or** thiocyanate; SNC^- carbidosulfidonitrate(1−)	SCN^- nitridosulfidocarbonato(1−) **or** thiocyanato; SNC^- carbidosulfidonitrato(1−)

Formula	Name	Cation	Anion	Additive anion name
COS	C(O)S carbonyl sulfide **or** oxidosulfidocarbon			
CO_2	CO_2 carbon dioxide **or** dioxidocarbon		$CO_2^{•-}$ oxidooxomethyl **or** dioxidocarbonate(•1−)	CO_2 dioxidocarbon
CO_3			CO_3^{2-} carbonate **or** trioxidocarbonate(2−)	CO_3^{2-} carbonato **or** trioxidocarbonato(2−)
CS	CS carbon monosulfide	$CS^{•+}$ sulfidocarbon(•1+)	$CS^{•-}$ sulfidocarbonate(•1−)	CS thiocarbonyl **or** sulfidocarbonato **or** carbon monosulfide
CS_2	CS_2 carbon disulfide **or** disulfidocarbon		$CS_2^{•-}$ sulfidothioxomethyl **or** disulfidocarbonate(•1−)	CS_2 disulfidocarbon
C_2	C_2 dicarbon	$C_2^{•+}$ dicarbon(•1+)	C_2^{2-} dicarbide(2−) **or** ethynediide **or** acetylenediide **or** acetylide	C_2^{2-} dicarbido **or** dicarbido(2−) **or** ethynediido **or** ethyne-1,2-diyl
Ca	calcium	calcium (general) Ca^{2+} calcium(2+)	calcide	calcido
Cd	cadmium	cadmium (general) Cd^{2+} cadmium(2+)	cadmide	cadmido
Cl	chlorine (general) $Cl^•$ chlorine(•) **or** monochlorine –Cl chloro	chlorine (general) Cl^+ chlorine(1+)	chloride (general) Cl^- chloride(1−) **or** chloride	chlorido (general) Cl^- chlorido(1−) **or** chlorido
ClF	ClF fluoridochlorine **or** chlorine monofluoride	$ClF^{•+}$ fluoridochlorine(•1+)		
ClF_2			ClF_2^- difluoridochlorate(1−)	ClF_2^- difluoridochlorato(1−)
ClF_4		ClF_4^+ tetrafluoridochlorine(1+)	ClF_4^- tetrafluoridochlorate(1−)	ClF_4^- tetrafluoridochlorato(1−)
ClH, see HCl				
ClHO	HOCl chloranol **or** hydroxidochlorine **or** hypochlorous acid		$HOCl^{•-}$ hydroxidochlorate(•1−)	
$ClHO_2$	HOClO hydroxy-λ^3-chloranone **or** hydroxidooxidochlorine **or** chlorous acid			
$ClHO_3$	$HOClO_2$ hydroxy-λ^5-chloranedione **or** hydroxidodioxidochlorine **or** chloric acid			
$ClHO_4$	$HOClO_3$ hydroxy-λ^7-chloranetrione **or** hydroxidotrioxidochlorine **or** perchloric acid			
ClNO	NOCl nitrogen chloride oxide **or** choridooxidonitrogen **or** nitrosyl chloride			

Table P7 *(Continued)*

Formula for uncharged atom or group	Name			
	Uncharged atoms or molecules (including zwitterions and radicals) or substituent groups[a]	Cations (including cation radicals) or cationic substituent groups[a]	Anions (including anion radicals) or anionic substituent groups[b]	Ligands[c]
$ClNO_2$	NO_2Cl nitrogen chloride dioxide **or** choridodioxidonitrogen **or** nitryl chloride			
ClO	OCl oxygen monochloride $OCl^{\bullet}$ chloridooxygen($\bullet$) **or** chlorosyl $-ClO$ oxo-λ^3-chloranyl **or** chlorosyl $-OCl$ chlorooxy		ClO^- chloridooxygenate(1–) **or** oxidochlorate(1–) **or** hypochlorite	ClO^- chloridooxygenato(1–) **or** oxidochlorato(1–) **or** hypochlorito
ClO_2	O_2Cl dioxygen chloride $ClO_2^{\bullet}$ dioxidochlorine($\bullet$) $ClOO^{\bullet}$ chloridodioxygen($O{-}O$)($\bullet$) $-ClO_2$ dioxo-λ^5-chloranyl **or** chloryl $-OClO$ oxo-λ^3-chloranyloxy	ClO_2^+ dioxidochlorine(1+) (**not** chloryl)	ClO_2^- dioxidochlorate(1–) **or** chlorite	ClO_2^- dioxidochlorato(1–) **or** chlorito
ClO_3	O_3Cl trioxygen chloride $ClO_3^{\bullet}$ trioxidochlorine($\bullet$) $-ClO_3$ trioxo-λ^7-chloranyl **or** perchloryl $-OClO_2$ dioxo-λ^5-chloranyloxy	ClO_3^+ trioxidochlorine(1+) (*not* perchloryl)	ClO_3^- trioxidochlorate(1–) **or** chlorate	ClO_3^- trioxidochlorato(1–) **or** chlorato
ClO_4	O_4Cl tetraoxygen chloride $ClO_4^{\bullet}$ tetraoxidochlorine($\bullet$) $-OClO_3$ trioxo-λ^7-chloranyloxy		ClO_4^- tetraoxidochlorate(1–) **or** perchlorate	ClO_4^- tetraoxidochlorato(1–) **or** perchlorato
Cl_2	Cl_2 dichlorine	$Cl_2^{\bullet+}$ dichlorine($\bullet$1+)	$Cl_2^{\bullet-}$ dichloride($\bullet$1–)	Cl_2 dichlorine $Cl_2^{\bullet-}$ dichlorido($\bullet$1–)
Cl_2OS	$SOCl_2$ sulfur dichloride oxide **or** dichloridooxidosulfur **or** sulfinyl dichloride **or** thionyl dichloride			
Cl_2O_2S	SO_2Cl_2 sulfur dichloride dioxide **or** dichloridodioxidosulfur **or** sulfonyl dichloride **or** sulfuryl dichloride			
Cl_3OP	$POCl_3$ phosphorus trichloride oxide **or** trichloridooxidophosphorus **or** phosphoryl trichloride			

Co	cobalt	cobalt (general) Co^{2+} cobalt(2+) Co^{3+} cobalt(3+)	cobaltide	cobaltido
Cr	chromium	chromium (general) Cr^{2+} chromium(2+) Cr^{3+} chromium(3+)	chromide	chromido
CrO_2	CrO_2 chromium dioxide **or** chromium(IV) oxide			
CrO_4	$[Cr(O_2)_2]$ diperoxidochromium		CrO_4^{2-} tetraoxidochromate(2−) **or** chromate CrO_4^{3-} tetraoxidochromate(3−) CrO_4^{4-} tetraoxidochromate(4−)	CrO_4^{2-} tetraoxidochromato(2−) **or** chromato CrO_4^{3-} tetraoxidochromato(3−) CrO_4^{4-} tetraoxidochromato(4−)
Cr_2O_3	Cr_2O_3 dichromium trioxide **or** chromium(III) oxide			
Cr_2O_7			$Cr_2O_7^{2-}$ heptaoxidodichromate(2−) $O_3CrOCrO_3^{2-}$ μ-oxido-bis(trioxido▲ chromate)(2−) **or** dichromate	$Cr_2O_7^{2-}$ heptaoxidodichromato(2−) $O_3CrOCrO_3^{2-}$ μ-oxido-bis(trioxido▲ chromato)(2−) **or** dichromato
Cu	copper	copper (general) Cu^{+} copper(1+) Cu^{2+} copper(2+)	cupride	cuprido
$D = {}^2H$	${}^2H^{\bullet} = D^{\bullet}$ deuterium(•) **or** monodeuterium	${}^2H^{+} = D^{+}$ deuterium(1+) **or** deuteron	${}^2H^{-} = D^{-}$ deuteride	deuterido
D_2	D_2 dideuterium	$D_2^{\bullet+}$ dideuterium(•1+)		
D_2O	$D_2O = {}^2H_2O$ dideuterium oxide **or** $({}^2H_2)$water			
FH, see HF				
F_2	F_2 difluorine	$F_2^{\bullet+}$ difluorine(•1+)	F_2 difluoride(•1−)	F_2 difluorine
For all F_nO_m see corresponding O_mF_n				
Fe	iron	iron (general) Fe^{2+} iron(2+) Fe^{3+} iron(3+)	ferride	ferrido

Table P7 *(Continued)*

Formula for uncharged atom or group	Name			
	Uncharged atoms or molecules (including zwitterions and radicals) or substituent groups[a]	Cations (including cation radicals) or cationic substituent groups[a]	Anions (including anion radicals) or anionic substituent groups[b]	Ligands[c]
GaH_3	GaH_3 gallium trihydride **or** gallane (parent hydride name) **or** trihydridogallium			
Ge	germanium $>Ge<$ germanetetrayl $=Ge=$ germanediylidene	germanium (general) Ge^{2+} germanium(2+) Ge^{4+} germanium(4+)	germide (general) Ge^{4-} germide(4−) **or** germide	germido (general) Ge^{4-} germido(4−) **or** germido
GeH_3	$-GeH_3$ germyl	GeH_3^+ germylium **or** trihydridogermanium(1+)	GeH_3^- germanide **or** trihydridogermanate(1−)	GeH_3 germanido **or** trihydridogermanato(1−)
GeH_4	GeH_4 germane (parent hydride name) **or** tetrahydridogermanium			
H, see also D and T	hydrogen $H^\bullet$ hydrogen(•) **or** monohydrogen (natural or unspecified isotopic composition) $^1H^\bullet$ protium(•) **or** monoprotium	hydrogen (general) H^+ hydrogen(1+) **or** hydron (natural or unspecified isotopic composition) $^1H^+$ protium(1+) **or** proton	hydride (general) H^- hydride (natural or unspecified isotopic composition) $^1H^-$ protide	hydrido protido
HBr	HBr hydrogen bromide **or** bromane (parent hydride name)			
HCl	HCl hydrogen chloride **or** chlorane (parent hydride name)	HCl^+ chloraniumyl **or** chloridohydrogen(•1+)		
HClO, see ClHO				
HF	HF hydrogen fluoride **or** fluorane (parent hydride name)	HF^+ fluoraniumyl **or** fluoridohydrogen(•1+)		
HF_2			FHF^- μ-hydridodifluorate(1−) **or** difluoridohydrogenate(1−)	
HI	HI hydrogen iodide **or** iodane (parent hydride name)			

Formula				
HIO	HOI iodanol **or** hydroxidoiodine **or** hypoiodous acid			
HIO_3	$HOIO_2$ hydroxy-λ^5-iodanedione **or** hydroxidodioxidoiodine **or** iodic acid		$HOIO_2^{\bullet-}$ hydroxidodioxidoiodate(•1−)	
HIO_4	$HOIO_3$ hydroxy-λ^7-iodanetrione **or** hydroxidotrioxidoiodine **or** periodic acid			
$HMnO_4$	$MnO_3(OH)$ hydroxidotrioxidomanganese		$MnO_3(OH)^-$ hydroxidotrioxidomanganate(1−)	
For all H_mN_m see corresponding N_nH_m				
HNO_2	$NO(OH)$ hydroxidooxidonitrogen **or** nitrous acid $>N(O)(OH)$ hydroxyoxo-λ^5-azanediyl **or** hydroxyazoryl $=N(O)(OH)$ hydroxyoxo-λ^5-azanylidene			
HNO_3	$NO_2(OH)$ hydroxidodioxidonitrogen **or** nitric acid $NO(OOH)$ (dioxidanido)oxidonitrogen **or** peroxynitrous acid			
HO	$HO^{\bullet}$ oxidanyl **or** hydridooxygen(•) **or** hydroxyl $–OH$ oxidanyl **or** hydroxy	HO^+ oxidanylium **or** hydridooxygen(1+) **or** hydroxylium	HO^- oxidanide **or** hydridooxygenate(1−) **or** hydroxide	HO^- oxidanido **or** hydroxido
HO_2	$HO_2^{\bullet}$ dioxidanyl **or** hydridodioxygen(•) **or** $–OOH$ dioxidanyl **or** hydroperoxy	HO_2^+ dioxidanylium **or** hydridodioxygen(1+)	HO_2^- dioxidanide **or** hydrogen(peroxide)(1−)	HO_2^- dioxidanido **or** hydrogen(peroxido)(1−)
HO_3P	HPO_3 metaphosphoric acid $(–P(O)(OH)O–)_n$ catena-poly▸ (hydroxidooxidophosphorus-μ-oxido)		$PO_2(OH)_2^{2-}$ hydroxidodioxidophosphate(2−) HPO_3^{2-} hydrogenphosphite PHO_3^{2-} hydridotrioxidophosphate(2−) **or** phosphonate	$PO_2(OH)_2^{2-}$ hydroxidodioxidophosphato(2−) HPO_3^{2-} hydrogenphosphito PHO_3^{2-} hydridotrioxidophosphato(2−) **or** phosphonato
HO_3S	$–S(O)_2(OH)$ hydroxydioxo-λ^6-sulfanyl **or** hydroxysulfonyl **or** sulfo		HSO_3^- hydroxidodioxidosulfate(1−) **or** hydrogensulfite	HSO_3^- hydroxidodioxidosulfato(1−) **or** hydrogensulfito

Table P7 (*Continued*)

Formula for uncharged atom or group	Name			
	Uncharged atoms or molecules (including zwitterions and radicals) or substituent groups[a]	Cations (including cation radicals) or cationic substituent groups[a]	Anions (including anion radicals) or anionic substituent groups[b]	Ligands[c]
HO_4P			$PO_3(OH)^{\bullet-}$ hydroxidotrioxidophosphate($\bullet$1−) $PO_3(OH)^{2-}$ hydroxidotrioxido▲ phosphate(2−) **or** hydrogenphosphate	$PO_3(OH)^{\bullet-}$ hydroxidotrioxidophosphato($\bullet$1−) $PO_3(OH)^{2-}$ hydroxidotrioxido▲ phosphato(2−) **or** hydrogenphosphato
HO_4S	$HOSO_3^{\bullet}$ hydroxidotrioxidosulfur($\bullet$) $-OS(O)_2(OH)$ hydroxysulfonyloxy **or** sulfooxy		HSO_4^{-} hydroxidotrioxidosulfate(1−) **or** hydrogensulfate	HSO_4 hydroxidotrioxidosulfato(1−) **or** hydrogensulfato
HS	$-SH$ sulfanyl $HS^{\bullet}$ sulfanyl **or** hydridosulfur($\bullet$)	HS^{+} sulfanylium **or** hydridosulfur(1+)	HS^{-} sulfanide **or** hydrogen(sulfide)(1−)	HS^{-} sulfanido **or** hydrogen(sulfido)(1−)
HS_2	$-SSH$ disulfanyl		HSS^{-} disulfanide	HSS^{-} disulfanido
HSe	$HSe^{\bullet}$ selanyl **or** hydridoselenium($\bullet$) $-SeH$ selanyl	HSe^{+} selanylium **or** hydridoselenium(1+)	HSe^{-} selanide **or** hydrogen(selenide)(1−)	HSe^{-} selanido **or** hydrogen(selenido)(1−)
HSe_2	$-SeSeH$ diselanyl		$HSeSe^{-}$ diselanide	$HSeSe^{-}$ diselanido
HTe	$HTe^{\bullet}$ tellanyl **or** hydridotellurium($\bullet$) $-TeH$ tellanyl	HTe^{+} tellanylium **or** hydridotellurium(1+)	HTe^{-} tellanide **or** hydrogen(tellanide)(1−)	
HTe_2	$-TeTeH$ ditellanyl		$HTeTe^{-}$ ditellanide	$HTeTe^{-}$ ditellanido
H_2	H_2 dihydrogen	$H_2^{\bullet+}$ dihydrogen($\bullet$1+) $^1H_2^{\bullet+}$ diprotium($\bullet$1+)		
H_2Br	$H_2Br^{\bullet}$ λ^3-bromanyl **or** dihydridobromine($\bullet$)	H_2Br^{+} bromanium **or** dihydridobromine(1+)		
H_2Cl	$H_2Cl^{\bullet}$ λ^3-chloranyl **or** dihydridochlorine($\bullet$)	H_2Cl^{+} chloranium **or** dihydridochlorine(1+)		
H_2F	$H_2F^{\bullet}$ λ^3-fluoranyl **or** dihydridofluorine($\bullet$)	H_2F^{+} fluoranium **or** dihydridofluorine(1+)		
H_2I	$H_2I^{\bullet}$ λ^3-iodanyl **or** dihydridoiodine($\bullet$)	H_2I^{+} iodanium **or** dihydridoiodine(1+)		
H_2MnO_4	$MnO_2(OH)_2$ dihydroxidodioxidomanganese			

Formula				
H₂NO	H₂NO• aminooxidanyl or dihydridooxidonitrogen(•) or aminoxyl / HONH• hydroxyazanyl or hydridohydroxidonitrogen(•) / –NH(OH) hydroxyazanyl or hydroxyamino / –ONH₂ aminooxy / –NH₂(O) oxo-λ⁵-azanyl or azinoyl		HONH⁻ hydroxyazanide or hydroxyhydroxidonitrate(1−) / H₂NO⁻ azanolate or aminooxidanide or dihydridooxidonitrate(1−)	NHOH⁻ hydroxyazanido or hydridohydroxidonitrato(1−) / H₂NO⁻ azanolato or aminooxidanido or dihydridooxidonitrato(1−)
H₂O	H₂O dihydrogen oxide or water or oxidane (parent hydride name) or dihydridooxygen / ¹H₂O diprotium oxide or (¹H₂)water			H₂O aqua
H₂OP	–PH₂O oxo-λ⁵-phosphanyl or phosphinoyl		PH₂O⁻ dihydridooxidophosphate(1−) or phosphinite	PH₂O⁻ dihydridooxidophosphato(1−) or phosphinito
H₂O₂	HOOH dihydrogen peroxide or hydrogen peroxide or dioxidane (parent hydride name) or bis(hydridooxygen)(O—O)	HOOH•⁺ dioxidaniumyl or bis(hydridooxygen)(O—O)(•1+)		HOOH dioxidane
H₂O₂P	–P(OH)₂ dihydroxyphosphanyl / –PH(O)(OH) hydroxyoxo-λ⁵-phosphanyl		PH₂O₂⁻ dihydridodioxido▲ phosphate(1−) or phosphinate	PH₂O₂⁻ dihydridodioxido▲ phosphato(1−) or phosphinato
H₂O₃P	–P(O)(OH)₂ dihydroxyoxo-λ⁵-phosphanyl or dihydroxyphosphoryl or phosphono		[PHO₂(OH)]⁻ hydridohydroxido▲ dioxidophosphate(1−) or hydrogenphosphonate / [PO(OH)₂]⁻ dihydridooxido▲ phosphate(1−) or dihydrogenphosphite	[PHO₂(OH)]⁻ hydridohydroxido▲ dioxidophosphato(1−) or hydrogenphosphonato / [PO(OH)₂]⁻ dihydridooxido▲ phosphato(1−) or dihydrogenphosphito
H₂O₄P	(HO)₂PO•₂ (dihydroxido)dioxido▲ phosphorus(•)		H₂PO₄ dihydroxidodioxido▲ phosphate(1−) or dihydrogenphosphate	H₂PO₄⁻ dihydroxidodioxido▲ phosphato(1−) or dihydrogenphosphato
H₂O₅P₂			[PH(O)₂OPH(O)₂]²⁻ μ-oxido-bis(hydridodioxidophosphate)(2−) or diphosphonate	[PH(O)₂OPH(O)₂]²⁻ μ-oxido-bis(hydridodioxidophosphato)(2−) or diphosphonato
H₂O₇S₂	[HOS(O)₂OS(O)₂OH] μ-oxido-bis▲ (hydroxidodioxidosulfur) or disulfuric acid			
H₂O₈S₂	[(HO)S(O)₂OOS(O)₂(OH)] μ-peroxido-1κO,2κO'-bis(hydroxidodioxidosulfur) or peroxydisulfuric acid			

Table P7 (*Continued*)

Formula for uncharged atom or group	Name			
	Uncharged atoms or molecules (including zwitterions and radicals) or substituent groups[a]	Cations (including cation radicals) or cationic substituent groups[a]	Anions (including anion radicals) or anionic substituent groups[b]	Ligands[c]
H_2S	H_2S dihydrogen sulfide **or** hydrogen sulfide **or** sulfane (parent hydride name) **or** dihydridosulfur	$H_2S^{•+}$ sulfaniumyl **or** dihydridosulfur(•1+)	$H_2S^{•-}$ sulfanuidyl **or** dihydridosulfate(•1−)	H_2S sulfane
H_2S_2	H_2S_2 dihydrogen disulfide HSSH disulfane (parent hydride name) **or** bis(hydridosulfur)(S—S)	$HSSH^{•+}$ disulfaniumyl **or** bis(hydridosulfur)(S—S)(•1+)	$HSSH^{•-}$ **or** disulfanuidyl **or** bis(hydridosulfate)(S—S)(•1−)	HSSH disulfane
H_2Se	H_2Se dihydrogen selenide **or** hydrogen selenide **or** selane (parent hydride name) **or** dihydridoselenium	$H_2Se^{•+}$ selaniumyl **or** dihydridoselenium(•1+)	$H_2Se^{•-}$ selanuidyl **or** dihydridoselenate(•1−)	H_2Se selane
H_2Se_2	H_2Se_2 dihydrogen diselenide HSeSeH diselane (parent hydride name) **or** bis(hydridoselenium)(Se—Se)	$HSeSeH^{•+}$ diselaniumyl **or** bis(hydridoselenium) (Se—Se)(•1+)	$HSeSeH^{•-}$ diselanuidyl **or** bis(hydridoselenate)(Se—Se)(•1−)	HSeSeH diselane
H_3NO	$HONH_2$ azanol **or** dihydridohydroxido▶ nitrogen **or** hydroxylamine (parent name for organic derivatives)	$HONH_2^{•+}$ hydroxyazaniumyl **or** dihydridohydroxido-nitrogen(•1+)		$HONH_2$ azanol **or** dihydridohydroxidonitrogen **or** hydroxylamine
H_3O		H_3O^+ oxidanium **or** trihydridooxygen(1+) **or** aquahydrogen(1+) **or** oxonium (**not** hydronium)		
H_3O_3P	$P(OH)_3$ trihydroxidophosphorus **or** phosphorous acid[e] $PHO(OH)_2$ hydridodihydroxidooxido▶ phosphorus **or** phosphonic acid[e]			
H_3O_4P	$PO(OH)_3$ trihydroxidooxidophosphorus **or** phosphoric acid			
H_3S	$H_3S^•$ λ^4-sulfanyl **or** trihydridosulfur(•)	H_3S^+ sulfanium **or** trihydridosulfur(1+)	H_3S^- sulfanuide **or** trihydridosulfate(1−)	
H_3Se	$H_3Se^•$ λ^4-selanyl **or** trihydridoselenium(•)	H_3Se^+ selanium **or** trihydridoselenium(1+)	H_3Se^- selanuide **or** trihydridoselenate(1−)	

216

Formula	Substituent group / neutral	Cation	Anion	Ligand
H_5IO_6	$IO(OH)_5$ pentahydroxy-λ^7-iodanone **or** pentahydroxidooxidoiodine **or** orthoperiodic acid			
H_4O_4Si	$Si(OH)_4$ silanetetrol **or** tetrahydroxidosilicon **or** silicic acid			
$H_4O_7P_2$	$[(HO)_2P(O)OP(O)(OH)_2]$ μ-oxido-bis(dihydroxidooxidophosphorus) **or** diphosphoric acid			
H_5O_2		$[H(H_2O)_2]^+$ μ-hydrido▶ bis(dihydridooxygen)(1+) **or** diaquahydrogen(1+)		
Hg	mercury	mercury (general) Hg^{2+} mercury(2+)	mercury (general) Hg^{2+} mercury(2+)	mercurido
Hg_2		Hg_2^{2+} dimercury(2+)		
I	iodine (general) $I^\bullet$ iodine($\bullet$) **or** monoiodine $–I$ iodo	iodine (general) I^+ iodine(1+)	iodide (general) I^- iodide(1–) **or** iodide	I^- iodido(1–) **or** iodido
ICl_2	$ICl_2^\bullet$ dichloridoiodine($\bullet$) $–ICl_2$ dichloro-λ^3-iodanyl	ICl_2^+ dichloroiodanium **or** dichloridoiodine(1+)		
IF	IF iodine fluoride **or** fluoridoiodine			
IF_4		IF_4^+ tetrafluoro-λ^3-iodanium **or** tetrafluoridoiodine(1+)	IF_4^- tetrafluoro-λ^3-iodanuide **or** tetrafluoridoiodate(1–)	IF_4^- tetrafluoridoiodato(1–)
IO	OI oxygen iodide $IO^\bullet$ iodidooxygen($\bullet$) **or** iodosyl $–IO$ oxo-λ^3-iodanyl **or** iodosyl $–OI$ iodooxy	IO^+ oxidoiodine(1+) (**not** iodosyl)	IO^- oxidoiodate(1–) **or** hypoiodite $IO^{\bullet 2-}$ oxidoiodate($\bullet$2–)	IO^- oxidoiodato(1–) **or** hypoiodito
IO_4	O_4I tetraoxygen iodide $IO_4^\bullet$ tetraoxidoiodine($\bullet$) $–OIO_3$ trioxo-λ^7-iodanyloxy		IO_4^- tetraoxidoiodate(1–) **or** periodate	IO_4^- tetraoxidoiodato(1–) **or** periodato
IO_6			IO_6^{5-} hexaoxidoiodate(5–) **or** orthoperiodate	IO_6^{5-} hexaoxidoiodato(5–) **or** orthoperiodato
I_2	I_2 diiodine	$I_2^{\bullet +}$ diiodine($\bullet$1+)	$I_2^{\bullet -}$ diiodide($\bullet$1–)	I_2 diiodine

Table P7 (*Continued*)

Formula for uncharged atom or group	Name			
	Uncharged atoms or molecules (including zwitterions and radicals) or substituent groups[a]	Cations (including cation radicals) or cationic substituent groups[a]	Anions (including anion radicals) or anionic substituent groups[b]	Ligands[c]
I_3	triodine		I_3^- triiodide(•1−) **or** triiodide	I_3^- triiodido(1−) **or** triiodido
InH_3	indium trihydride **or** indigane (parent hydride name) **or** trihydridoindium			
K	potassium	potassium	potasside	potassido
KO_2	potassium dioxide(1−) **or** potassium superoxide			
K_2O	dipotassium oxide			
K_2O_2	dipotassium dioxide(2−) **or** potassium peroxide			
La	lanthanum	lanthanum	lanthanide[d]	lanthanido
Li	lithium	lithium (general) Li^+ lithium(1+)	lithide (general) Li^- lithide(1−) **or** lithide	lithido Li^- lithido(1−) **or** lithido
$LiCl$	lithium chloride	$LiCl^+$ chloridolithium(1+)	$LiCl^-$ chloridolithate(1−)	$LiCl^-$ chloridolithato(1−)
LiH	lithium hydride **or** hydridolithium	LiH^+ hydridolithium(1+)	LiH^- hydridolithate(1−)	LiH^- hydridolithato(1−)
Li_2	dilithium	Li_2^+ dilithium(•1+)	Li_2^- dilithide(•1−)	Li_2^- dilithido(•1−)
Mg	magnesium	magnesium (general) Mg^{2+} magnesium(2+)	magneside (general) Mg^- magneside(1−)	magnesido Mg^- magnesido(1−)
Mn	manganese	manganese (general) Mn^{2+} manganese(2+) Mn^{3+} manganese(3+)	manganide	manganido
MnO	manganese mon(o)oxide **or** manganese(II) oxide			
MnO_2	manganese dioxide **or** manganese(IV) oxide			
MnO_3		MnO_3^+ trioxidomanganese(1+)		

MnO_4			MnO_4^- tetraoxidomanganate(1−) **or** permanganate MnO_4^{2-} tetraoxidomanganate(2−) **or** manganate(VI)	MnO_4^- tetraoxidomanganato(1−) **or** permanganato MnO_4^{2-} tetraoxidomanganato(2−) **or** manganato(VI)
Mn_2O_7	Mn_2O_7 dimanganese heptaoxide **or** manganese(VII) oxide **or** $O_3MnOMnO_3$ μ-oxidobis(trioxidomanganese)			
Mo	molybdenum	molybdenum	molybdenide	molybdenido
N	nitrogen $N^{\bullet\bullet\bullet}$ nitrogen(•) **or** mononitrogen $-N<$ azanetriyl **or** nitrilo $-N=$ azanylylidene $\equiv N$ azanylidyne	nitrogen (general) N^+ nitrogen(1+)	nitride (general) N^{3-} nitride(3−) **or** azanetriide **or** nitride $=N^-$ azanidylidene **or** amidylidene $-N^{2-}$ azanediidyl	N^{3-} nitrido(3−) **or** azanetriido
NF_4		NF_4^+ tetrafluoroammonium **or** tetrafluoroazanium **or** tetrafluoridonitrogen(1+)		
NH	$NH^{\bullet\bullet}$ azanylidene **or** hydridonitrogen(2•) **or** nitrene $>NH$ azanediyl $=NH$ azanylidene **or** imino	$NH^{\bullet+}$ azanyliumyl **or** hydridonitrogen(•1+) NH^{2+} azanebis(ylium) **or** hydridonitrogen(2+)	NH^- azanidyl **or** hydridonitrate(•1−) NH^{2-} azanediide **or** hydridonitrate(2−) **or** imide $-NH^-$ azanidyl **or** amidyl	NH^{2-} azanediido **or** hydridonitrato(2−) **or** imido
NH_2	$NH_2^\bullet$ azanyl **or** dihydridonitrogen(•) **or** aminyl $-NH_2$ azanyl **or** amino	NH_2^+ azanylium **or** dihydridonitrogen(1+)	NH_2^- azanide **or** dihydridonitrate(1−) **or** amide	NH_2^- azanido **or** dihydridonitrato(1−) **or** amido
NH_3	NH_3 azane (parent hydride name) **or** amine (parent name for certain organic derivatives) **or** trihydridonitrogen **or** ammonia	$NH_3^{\bullet+}$ azaniumyl **or** trihydridonitrogen(•1+) $-NH_3^+$ azaniumyl **or** ammonio	NH_3^- azanuidyl **or** trihydridonitrate(•1−)	NH_3 ammine
NH_4	$NH_4^\bullet$ λ^5-azanyl **or** tetrahydridonitrogen(•)	NH_4^+ azanium **or** ammonium		
NO	NO nitrogen mon(o)oxide (**not** nitric oxide) $NO^\bullet$ oxoazanyl **or** oxidonitrogen(•) **or** nitrosyl $-N=O$ oxoazanyl **or** nitroso $>N(O)-$ oxo-λ^5-azantriyl **or** azoryl **or** nitroryl $=N(O)-$ oxo-λ^5-azanylylidene	NO^+ oxidonitrogen(1+) (**not** nitrosyl) $NO^{\bullet2+}$ oxidonitrogen(•2+)	NO^- oxidonitrate(1−)	NO oxidonitrogen (general) nitrosyl = oxidonitrogen-κN NO^+ oxidonitrogen(1+) NO^- oxidonitrato(1−)

Table P7 *(Continued)*

Formula for uncharged atom or group	Name			
	Uncharged atoms or molecules (including zwitterions and radicals) or substituent groups[a]	Cations (including cation radicals) or cationic substituent groups[a]	Anions (including anion radicals) or anionic substituent groups[b]	Ligands[c]
NO_2	NO_2 nitrogen dioxide $ONO^\bullet$ nitrosooxidanyl **or** dioxidonitrogen($\bullet$) **or** nitryl $-NO_2$ nitro $-ONO$ nitrosooxy	NO_2^+ dioxidonitrogen(1+) (**not** nitryl)	NO_2^- dioxidonitrate(1−) **or** nitrite $NO_2^{\bullet 2-}$ dioxidonitrate($\bullet$2−)	NO_2^- dioxidonitrato(1−) **or** nitrito $NO_2^{\bullet 2-}$ dioxidonitrato($\bullet$2−)
NO_3	NO_3 nitrogen trioxide $O_2NO^\bullet$ nitrooxidanyl **or** trioxidonitrogen($\bullet$) **or** $ONOO^\bullet$ nitrosodioxidanyl **or** oxidoperoxidonitrogen($\bullet$) $-ONO_2$ nitrooxy		NO_3^- trioxidonitrate(1−) **or** nitrate $NO_3^{\bullet 2-}$ trioxidonitrate($\bullet$2−) $[NO(OO)]^-$ oxidoperoxidonitrate(1−) **or** peroxynitrite	NO_3^- trioxidonitrato($\bullet$1−) **or** nitrato $NO_3^{\bullet 2-}$ trioxidonitrato($\bullet$2−) $[NO(OO)]^-$ oxidoperoxidonitrato(1−) **or** peroxynitrito
NO_4			$NO_2(O_2)^-$ dioxidoperoxidonitrate(1−) **or** peroxynitrate	$NO_2(O_2)^-$ dioxidoperoxidonitrato(1−) **or** peroxynitrato
NS	NS nitrogen monosulfide $NS^\bullet$ sulfidonitrogen($\bullet$) $-N=S$ sulfanylideneazanyl **or** thionitroso	NS^+ sulfidonitrogen(1+) (**not** thionitrosyl)	NS^- sulfidonitrate(1−)	NS thionitrosyl **or** sulfidonitrogen **or** sulfidonitrato (general) NS^+ sulfidonitrogen(1+) NS^- sulfidonitrato(1−)
N_2	N_2 dinitrogen $=N^+=N^-$ diazo **or** (azanidylidene)azaniumylidene $=N-N=$ diazane-1,2-diylidene **or** hydrazinediylidene $-N=N-$ diazene-1,2-diyl **or** azo	$N_2^{\bullet +}$ dinitrogen(1+) N_2^{2+} dinitrogen(2+)	N_2^{2-} dinitride(2−) N_2^{4-} dinitride(4−) **or** diazanetetraide **or** hydrazinetetraide	N_2 dinitrogen N_2^{2-} dinitrido(2−) N_2^{4-} dinitrido(4−) **or** diazanetetraido **or** hydrazinetetraido
N_2F_2	$FN=NF$ difluorido-1κF,2κF-dinitrogen(N–N) **or** difluorodiazene			
N_2H	$N\equiv NH^+$ diazynium	$N\equiv NH^+$ diazynium	$N=NH^-$ diazenide NNH^{3-} diazanetride **or** hydrazinetride	$N=NH^-$ diazenido NNH^{3-} diazanetriido **or** hydrazinetriido

Formula				
N₂H₂	HN=NH diazene ⁻N=NH₂⁺ diazen-2-ium-1-ide H₂NN²⁻• diazanylidene or hydrazinylidene =NNH₂ diazanylidene or hydrazinylidene •HNNH• diazane-1,2-diyl or hydrazine-1,2-diyl −HNNH− diazane-1,2-diyl or hydrazine-1,2-diyl	HNNH²⁺ diazynediium or diazanediylium or hydrazinediylium	HNNH²⁻ diazane-1,2-diide or hydrazine-1,2-diide or H₂NN²⁻ diazane-1,1-diide or hydrazine-1,1-diide	HN=NH diazene HNNH²⁻ diazane-1,2-diido or hydrazine-1,2-diido or H₂NN²⁻ diazane-1,1-diido or hydrazine-1,1-diido
N₂H₃	H₂N-NH• diazanyl or trihydridodi▲ nitrogen(N—N)(•) or hydrazinyl -NH-NH₂ diazanyl or hydrazinyl	H₂N=NH⁺ diazenium	H₂N-NH⁻ diazanide or hydrazinide ²⁻N-NH₃⁺ diazan-2-ium-1,1-diide	H₂N-NH⁻ diazanido or hydrazinido ²⁻N-NH₃⁺ diazan-2-ium-1,1-diido
N₂H₄	H₂N-NH₂ diazane (parent hydride name) or hydrazine (parent name for organic derivatives)	H₂N-NH₂•⁺ diazaniumyl or bis(dihydridonitrogen)▲ (N—N)(•1+) or hydraziniumyl H₂N=NH₂⁺ diazenediium ⁻NHNH₃⁺ diazan-2-ium-1-ide		H₂N-NH₂ diazane or hydrazine
N₂H₅		H₂N-NH₃⁺ diazanium or hydrazinium		
N₂H₆		H₃N-NH₃²⁺ diazanediium or hydrazinediium		
N₂O	N₂O dinitrogen oxide -N(O)=N- azoxy		N₂O•⁻ oxidodinitrate(•1-)	N₂O dinitrogen oxide (general) N₂O•⁻ oxidodinitrato(•1-)
N₂O₂	N₂O₂ dinitrogen dioxide ON-NO bis(oxidonitrogen)(N—N)		N₂O₂²⁻ = ONNO²⁻ diazenediolate or bis(oxidonitrate)(N—N)(2-)	N₂O₂²⁻ bis(oxidonitrato)(N—N)(2-)
N₂O₃	N₂O₃ dinitrogen trioxide O₂N-NO trioxido-1κ²O,2κO-▲ dinitrogen(N—N) NO⁺NO₂⁻ oxidonitrogen(1+)dioxidonitrate(1-)		O₂N-NO²⁻ trioxido-1κ²O,2κO-▲ dinitrate(N—N)(2-)	
N₂O₄	N₂O₄ dinitrogen tetraoxide O₂N-NO₂ bis(dioxidonitrogen)(N—N) NO⁺NO₃⁻ oxidonitrogen(1+)trioxidonitrate(1-)			
N₂O₅	N₂O₅ dinitrogen pentaoxide O₂NONO₂ dinitrooxidane or μ-oxido-bis(dioxidonitrogen)(N—N) NO₂⁺NO₃⁻ dioxidonitrogen(1+)trioxidonitrate(1-)			

Table P7 *(Continued)*

Formula for uncharged atom or group	Name			
	Uncharged atoms or molecules (including zwitterions and radicals) or substituent groups[a]	Cations (including cation radicals) or cationic substituent groups[a]	Anions (including anion radicals) or anionic substituent groups[b]	Ligands[c]
N_3	$N_3^{\bullet}$ trinitrogen($\bullet$) $-N=N^+=N^{\bullet}$ azido		N_3^- trinitride($1-$) **or** azide	N_3^- trinitrido($1-$) **or** azido
N_3H	N_3H hydrogen trinitride **or** hydrogen azide $HN=N^+=N^-$ hydrido-$1\kappa H$-trinitrogen ▶ ($2\,N-N$)			
Na	sodium	sodium (general) Na^+ sodium($1+$)	sodide (general) Na^- sodide($1-$) **or** sodide	sodido Na^- sodido($1-$) **or** sodido
Ni	nickel	nickel (general) Ni^{2+} nickel($2+$) Ni^{3+} nickel($3+$)	nickelide	nickelido
O	oxygen (general) O monooxygen $O^{2\bullet}$ oxidanylidene **or** monooxygen($2\bullet$) $>O$ oxy **or** epoxy (in rings) $=O$ oxo	oxygen (general) $O^{\bullet+}$ oxygen($\bullet 1+$)	oxide (general) $O^{\bullet-}$ oxidanidyl **or** oxide($\bullet 1-$) O^{2-} oxide($2-$) **or** oxide $-O^-$ oxido	oxido (general) O^{2-} oxido ($2-$)
OCl, see ClO				
OF	OF oxygen monofluoride $OF^{\bullet}$ fluoridooxygen($\bullet$) $-FO$ oxo-λ^3-fluoranyl **or** fluorosyl	OF^+ fluoridooxygen($1+$)	OF^- fluoridooxygenate($1-$)	
OF_2	OF_2 oxygen difluoride **or** difluoridooxygen			
O_2	O_2 dioxygen $O_2^{2\bullet}$ dioxidanediyl **or** dioxygen($2\bullet$) $-OO-$ dioxidanediyl **or** peroxy	$O_2^{\bullet+}$ dioxidanyliumyl **or** dioxygen($\bullet 1+$) O_2^{2+} dioxidanebis(ylium) **or** dioxygen($2+$)	$O_2^{\bullet-}$ dioxidanidyl **or** dioxide($\bullet 1-$) **or** superoxide (**not** hyperoxide) O_2^{2-} dioxidanediide **or** dioxide($2-$) **or** peroxide	dioxido (general) O_2 dioxygen $O_2^{\bullet-}$ dioxido($\bullet 1-$) **or** superoxido O_2^{2-} dioxidanediido **or** dioxido($2-$) **or** peroxido
O_2Cl, see ClO_2				

O_2F_2	O_2F_2 dioxygen difluoride FOOF difluorodioxidane **or** bis(fluoridooxygen)(O—O)			
O_3Cl, see ClO_3				
O_4Cl, see ClO_4				
O_3	O_3 trioxygen **or** ozone $-OOO-$ trioxidanediyl		$O_3^{\bullet-}$ trioxide(•1–) **or** ozonide	O_3 trioxygen **or** ozone $O_3^{\bullet-}$ trioxido(•1–) **or** ozonido
Os	osmium	osmium	osmide	osmido
P	phosphorus (general) $P^{\bullet}$ phosphorus(•) **or** monophosphorus $>P-$ phosphanetriyl	phosphorus (general) P^+ phosphorus(1+)	phosphide (general) P^- phosphide(1–) P^{3-} phosphide(3–) **or** phosphanetriide **or** phosphide	P^{3-} phosphido **or** phosphanetriido
PF_2			PF_2^- difluorophosphanide **or** difluoridophosphate(1–)	PF_2^- difluorophosphanido **or** difluoridophosphato(1–)
PF_3	PF_3 phosphorus trifluoride **or** trifluorophosphane **or** trifluoridophosphorus			PF_3 trifluorophosphane **or** trifluoridophosphorus
PF_4		PF_4^+ tetrafluorophosphanium **or** tetrafluoridophosphorus(1+)	PF_4^- tetrafluorophosphanide **or** tetrafluoridophosphate(1–)	PF_4^- tetrafluorophosphanuido **or** tetrafluoridophosphato(1–)
PF_5	PF_5 phosphorus pentafluoride **or** pentafluoro-λ^5-phosphane **or** pentafluoridophosphorus			
PF_6			PF_6^- hexafluoro-λ^5-phosphanuide **or** hexafluoridophosphate(1–)	PF_6^- hexafluoro-λ^5-phosphanuido **or** hexafluoridophosphato(1–)
PH	$PH^{2\bullet}$ phosphanylidene **or** hydridophosphorus(2•) $>PH$ phosphanediyl $=PH$ phosphanylidene	PH^+ phosphanyliumyl **or** hydridophosphorus(1+) PH^{2+} phosphanebis(ylium) **or** hydridophosphorus(2+)	PH^+ phosphanidyl **or** hydridophosphate(1–) PH^{2-} phosphanediide **or** hydridophosphate(2–)	PH^{2-} phosphanediido **or** hydridophosphato(2–)
PH_2	$PH_2^{\bullet}$ phosphanyl **or** dihydridophosphorus(•) $-PH_2$ phosphanyl	PH_2^+ phosphanylium **or** dihydridophosphorus(1+)	PH_2^- phosphanide **or** dihydridophosphate(1–)	PH_2^- phosphanido **or** dihydridophosphato(1–)
PH_3	PH_3 phosphorus trihydride **or** phosphane (parent hydride name) **or** trihydridophosphorus	$PH_3^{\bullet+}$ phosphaniumyl **or** trihydridophosphorus(•1+) **or** $-PH_3$ phosphaniumyl	PH_3^- phosphanuidyl **or** trihydridophosphate(•1–)	PH_3 phosphane

Table P7 (*Continued*)

Formula for uncharged atom or group	Name			
	Uncharged atoms or molecules (including zwitterions and radicals) or substituent groups[a]	Cations (including cation radicals) or cationic substituent groups[a]	Anions (including anion radicals) or anionic substituent groups[b]	Ligands[c]
PH_4	$-PH_4$ λ^5-phosphanyl	PH_4^+ phosphanium **or** tetrahydridophosphorus(1+)	PH_4^- phosphanuide **or** tetrahydridophosphate(1−)	PH_4^- phosphanuido **or** tetrahydridophosphato(1−)
PH_5	PH_5 phosphorus pentahydride **or** λ^5-phosphane (parent hydride name) **or** pentahydridophosphorus			
PO	$PO^\bullet$ oxophosphanyl **or** oxidophosphorus(•) **or** phosphorus mon(o)oxide **or** phosphoryl; $>P(O)-$ oxo-λ^5-phosphanetriyl **or** phosphoryl; $=P(O)-$ oxo-λ^5-phosphanylidene; $\equiv PO$ oxo-λ^5-phosphanylidyne	PO^+ oxidophosphorus(1+) (**not** phosphoryl)	PO^- oxidophosphate(1−)	PO^- oxidophosphato(1−)
PO_2	$-PO_2$ dioxo-λ^5-phosphanyl		PO_2^- dioxidophosphate(1−)	PO_2^- dioxidophosphato(1−)
PO_3			PO_3^- trioxidophosphate(1−); $PO_3^{\bullet 2-}$ trioxidophosphate(•2−); PO_3^{3-} trioxidophosphate(3−) **or** phosphite; $(P(O)_2O)_n^{n-}$ *catena-*▶ poly[(dioxidophosphate-μ-oxido)(1−)] **or** metaphosphate; $-P(O)(O^-)_2$ dioxidooxo-λ^5-▶ phosphanyl **or** phosphonato	PO_3^- trioxidophosphato(1−); $PO_3^{\bullet 2-}$ trioxidophosphato(•2−); PO_3^{3-} trioxidophosphato(3−) **or** phosphito
PO_4			$PO_4^{\bullet 2-}$ tetraoxidophosphate(•2−); PO_4^{3-} tetraoxidophosphate(3−) **or** phosphate	PO_4^{3-} tetraoxidophosphato(3−) **or** phosphato
PS_4			PS_4^{3-} tetrasulfidophosphate(3−)	PS_4^{3-} tetrasulfidophosphato(3−)
P_2	P_2 diphosphorus	P_2^+ diphosphorus(1+)	P_2^- diphosphide(1−); P_2^{2-} diphosphide(2−)	P_2 diphosphorus; P_2^- diphosphido(1−); P_2^{2-} diphosphido(2−)

P_2H_3	$H_2PPH^\bullet$ diphosphanyl **or** trihydridodiphosphorus$(P{-}P)(\bullet)$ -$HPPH_2$ diphosphanyl		H_2PPH^- diphosphanide	H_2PPH^- diphosphanido
P_2H_4	H_2PPH_2 diphosphane (parent hydride name)			H_2PPH_2 diphosphane
P_2O_6			$O_3PPO_3^{4-}$ bis(trioxidophosphate)▲ $(P{-}P)(4-)$ **or** hypodiphosphate	$O_3PPO_3^{4-}$ bis(trioxidophosphato)▲ $(P{-}P)(4-)$ **or** hypodiphosphato
P_2O_7			$O_3POPO_3^{4-}$ μ-oxido-bis▲ (trioxidophosphato)(4−) **or** diphosphate	$O_3POPO_3^{4-}$ μ-oxido-bis▲ (trioxidophosphato)(4−) **or** diphosphato
P_3O_{10}			$[O_3POP(O)_2OPO_3]^{5-}$ μ-phosphato-1κO,2κO'-bis▲ (trioxidophosphato)(5−) **or** triphosphate	$[O_3POP(O)_2OPO_3]^{5-}$ μ-phosphato-1κO,2κO'-bis▲ (trioxidophosphato)(5−) **or** triphosphato
P_4	P_4 tetraphosphorus			P_4 tetraphosphorus
Pb	lead	lead (general) Pb^{2+} lead(2+) Pb^{4+} lead(4+)	plumbide	plumbido
PbH_4	PbH_4 plumbane (parent hydride name) **or** tetrahydridolead **or** lead tetrahydride			
Pb_9			Pb_9^{4-} nonaplumbide(4−)	
Pd	palladium	palladium (general) Pd^{2+} palladium(2+) Pd^{4+} palladium(4+)	palladide	palladido
Pt	platinum	platinum (general) Pt^{2+} platinum(2+) Pt^{4+} platinum(4+)	platinide	platinido
Pu	plutonium	plutonium	plutonide	plutonido
PuO_2	PuO_2 plutonium dioxide	PuO_2^+ dioxidoplutonium(1+) [**not** plutonyl(1+)] PuO_2^{2+} dioxidoplutonium(2+) [**not** plutonyl (2+)]		
Rb	rubidium	rubidium	rubidide	rubidido

Table P7 *(Continued)*

Formula for uncharged atom or group	Name			
	Uncharged atoms or molecules (including zwitterions and radicals) or substituent groups[a]	Cations (including cation radicals) or cationic substituent groups[a]	Anions (including anion radicals) or anionic substituent groups[b]	Ligands[c]
ReO_4			ReO_4^- tetraoxidorhenate(1−) ReO_4^{2-} tetraoxidorhenate(2−)	ReO_4^- tetraoxidorhenato(1−) ReO_4^{2-} tetraoxidorhenato(2−)
Ru	ruthenium	ruthenium	ruthenide	ruthenido
S	sulfur (general) S monosulfur =S sulfanylidene **or** thioxo −S− sulfanediyl	sulfur (general) $S^{\bullet+}$ sulfur(•1+)	sulfide (general) $S^{\bullet-}$ sulfanidyl **or** sulfide(•1−) S^{2-} sulfanediide **or** sulfide(2−) **or** sulfide −S− sulfido	sulfido (general) $S^{\bullet-}$ sulfido(•1−) S^{2-} sulfanediido **or** sulfido(2−) **or** sulfido
SCN, see CNS				
SH, see HS				
SH₂, see H₂S				
SNC, see CNS				
SO	SO sulfur mon(o)oxide **or** oxidosulfur >SO oxo-λ^4-sulfanediyl **or** sulfinyl **or** thionyl	$SO^{\bullet+}$ oxidosulfur(•1+) (**neither** sulfinyl **nor** thionyl)	$SO^{\bullet-}$ oxidosulfate(•1−)	SO oxidosulfur
SO_2	SO_2 sulfur dioxide **or** dioxidosulfur >SO_2 dioxo-λ^6-sulfanediyl **or** sulfonyl **or** sulfuryl		$SO_2^{\bullet-}$ dioxidosulfate(•1−) SO_2^{2-} dioxidosulfate(2−) **or** sulfanediolate	SO_2 dioxidosulfur SO_2^{2-} dioxidosulfato(2−) **or** sulfanediolato
SO_3	SO_3 sulfur trioxide		$SO_3^{\bullet-}$ trioxidosulfate(•1−) SO_3^{2-} trioxidosulfate(2−) **or** sulfite −S(O)$_2$(O$^-$) oxidodioxo-λ^6-sulfanyl **or** sulfonato	SO_3^{2-} trioxidosulfato(2−) **or** sulfito
SO_4	−OS(O)$_2$O− sulfonylbis(oxy)		$SO_4^{\bullet-}$ tetraoxidosulfate(•1−) SO_4^{2-} tetraoxidosulfate(2−) **or** sulfate	SO_4^{2-} tetraoxidosulfato(2−) **or** sulfato

		$S_2^{•+}$ disulfur(•1+)	$S_2^{•−}$ disulfanidyl **or** disulfide(•1−) $S_2^{2−}$ disulfide(2−) **or** disulfanediide −SS− disulfanidyl	disulfido (general) $S_2^{2−}$ disulfido(2−) **or** disulfanediido $S_2^{•−}$ disulfido(•1−)
S_2	S_2 disulfur −SS− disulfanediyl >S=S sulfanyliden-λ^4-sulfanediyl **or** sulfinothioyl			
S_2O	>S(=O)(=S) oxosulfanyliden-λ^6-sulfanediyl **or** sulfonothioyl			
S_2O_3			$SO_3S^{•−}$ trioxido-1κ^3O-disulfate(S−S) (•1−) **or** trioxidosulfidosulfate(•1−) $SO_3S^{2−}$ trioxido-1κ^3O-disulfate▲ (S−S)(2−) **or** trioxidosulfidosulfate(2−) **or** thiosulfate	$SO_3S^{2−}$ trioxido-1κ^3O-▲ disulfato(S−S)(2−) **or** trioxidosulfidosulfato(2−) **or** thiosulfato
S_2O_4			$O_2SSO_2^{2−}$ bis(dioxidosulfate)(S−S)(2−) **or** dithionite	$O_2SSO_2^{2−}$ bis(dioxidosulfato)▲ (S−S)(2−) **or** dithionito
S_2O_5			$O_3SSO_2^{2−}$ pentaoxido-▲ 1κ^3O,2κ^2O-disulfate(S−S)(2−)^f $O_2SOSO_2^{2−}$ μ-oxido-bis(dioxidosulfate)(2−)^f	$O_3SSO_2^{2−}$ pentaoxido-▲ 1κ^3O,2κ^2O-disulfato(S−S)(2−) $O_2SOSO_2^{2−}$ μ-oxido-bis(dioxidosulfato)(2−)
S_2O_6			$O_3SSO_3^{2−}$ bis(trioxidosulfate)▲ (S−S)(2−) **or** dithionate	$O_3SSO_3^{2−}$ bis(trioxidosulfato)▲ (S−S)(2−) **or** dithionato
S_2O_7			$O_3SOSO_3^{2−}$ μ-oxido-bis▲ (trioxidosulfate)(2−) **or** disulfate	$O_3SOSO_3^{2−}$ μ-oxido-bis▲ (trioxidosulfato)(2−) **or** disulfato
S_2O_8			$O_3SOOSO_3^{2−}$ μ-peroxido-1κO,2$\kappa O'$-▲ bis(trioxidosulfate)(2−) **or** peroxydisulfate	$O_3SOOSO_3^{2−}$ μ-peroxido-1κO,2$\kappa O'$-▲ bis(trioxidosulfato)(2−) **or** peroxydisulfato
S_3	S_3 trisulfur −SS− trisulfanediyl >S(=S)$_2$ bis(sulfanylidene)-λ^6-▲ sulfanediyl **or** sulfonodithioyl	S_3^{2+} trisulfur(2+)	$S_3^{•−}$ trisulfide(•1−) SSS$^{•−}$ trisulfanidyl $S_3^{2−}$ trisulfide(2−) SSS$^{2−}$ trisulfanediide	$S_3^{•−}$ trisulfido(•1−) SSS$^{•−}$ trisulfanidylo $S_3^{2−}$ trisulfido(2−) SSS$^{2−}$ trisulfanediido
S_4	S_4 tetrasulfur −SSSS− tetrasulfanediyl	S_4^{2+} tetrasulfur(2+)	$S_4^{2−}$ tetrasulfide(2−) SSSS$^{2−}$ tetrasulfanediide	$S_4^{2−}$ tetrasulfido(2−) SSSS$^{2−}$ tetrasulfanediido
S_4O_6			$O_3SSSSO_3^{2−}$ disulfanedisulfonate **or** bis[(trioxidosulfato)sulfate(S−S)](2−) **or** tetrathionate $O_3SSSSO_3^{•3−}$ bis[(trioxidosulfato)▲ sulfate](S−S)(•3−)	$O_3SSSSO_3^{2−}$ disulfanedisulfonato **or** bis[(trioxidosulfato)sulfato] (S−S)](2−) **or** tetrathionato $O_3SSSSO_3^{•3−}$ bis[(trioxidosulfato)▲ sulfato](S−S)](•3−)

Table P7 (*Continued*)

Formula for uncharged atom or group	Name			
	Uncharged atoms or molecules (including zwitterions and radicals) or substituent groups[a]	Cations (including cation radicals) or cationic substituent groups[a]	Anions (including anion radicals) or anionic substituent groups[b]	Ligands[c]
S_8	S_8 octasulfur	S_8^{2+} octasulfur(2+)	S_8^{2-} octasulfide(2−) $S[S]_6S^{2-}$ octasulfanediide	S_8 octasulfur S_8^{2-} octasulfido(2−) $S[S]_6S^{2-}$ octasulfanediido
Sb	antimony >Sb– stibanetriyl	antimony	antimonide (general) Sb^{3-} antimonide(3−) **or** stibanetriide	antimonido (general) Sb^{3-} antimonido(3−) **or** stibanetriido
SbH_3	SbH_3 antimony trihydride **or** stibane (parent hydride name) **or** trihydridoantimony	$SbH_3^{\bullet+}$ stibaniumyl **or** trihydridoantimony(•1+) –SbH_3^+ stibaniumyl	$SbH_3^{\bullet-}$ stibanuidyl **or** trihydridoantimonate(•1−)	SbH_3 stibane
SbH_4	–SbH_4 λ^5-stibanyl	SbH_4^+ stibanium **or** tetrahydridoantimony(1+)		
SbH_5	SbH_5 antimony pentahydride **or** λ^5-stibane (parent hydride name) **or** pentahydridoantimony			
Se	Se (general) Se monoselenium >Se selanediyl =Se selanylidene **or** selenoxo	selenium	selenide (general) $Se^{\bullet-}$ selanidyl **or** selenide(•1−) Se^{2-} selanediide **or** selenide(2−) **or** selenide	selenido (general) $Se^{\bullet-}$ selenido(•1−) Se^{2-} selanediido **or** selenido(2−)
SeH, see HSe				
SeH$_2$, see H$_2$Se				
SeO_3	SeO_3 selenium trioxide		$SeO_3^{\bullet-}$ trioxidoselenate(•1−) SeO_3^{2-} trioxidoselenate(2−) **or** selenite	$SeO_3^{\bullet-}$ trioxidoselenato(•1−) SeO_3^{2-} trioxidoselenato(2−) **or** selenito
SeO_4			SeO_4^{2-} tetraoxidoselenate(2−) **or** selenate	SeO_4^{2-} tetraoxidoselenato(2−) **or** selenato
Si	silicon >Si< silanetetrayl =Si= silanediylidene	silicon (general) Si^+ silicon(•1+) Si^{4+} silicon(4+)	silicide (general) Si^- silicide(•1−) Si^{4-} silicide(4−) **or** silicide	silicido (general) Si^- silicido(•1−) Si^{4-} silicido(4−) **or** silicido

Formula	Neutral / parent	Cation	Anion (–ide / –ate)	Anion (–ido / –ato)
SiC	SiC silicon carbide **or** carbidosilicon	SiC^+ carbidosilicon(1+)		
SiH		SiH^+ silanyliumdiyl **or** hydridosilicon(1+)	SiH^- silanidediyl **or** hydridosilicate(1−)	
SiH_2	$>SiH_2$ silanediyl $=SiH_2$ silylidene			
SiH_3	$SiH_3^{\bullet}$ silyl **or** trihydridosilicon($\bullet$) $-SiH_3$ silyl	SiH_3^+ silylium **or** trihydridosilicon(1+)	SiH_3^- silanide **or** trihydridosilicate(1−)	SiH_3^- silanido
SiH_4	SiH_4 silicon tetrahydride **or** silane (parent hydride name) **or** tetrahydridosilicon			
SiO	SiO oxidosilicon **or** silicon mon(o)oxide	$SiO^{\bullet+}$ oxidosilicon($\bullet$1+)		
SiO_2	SiO_2 silicon dioxide			
SiO_3			$SiO_3^{\bullet-}$ trioxidosilicate($\bullet$1−) $(Si(O)_2O)_n^{2n-}$ catena-poly▶ $[(dioxidosilicate-\mu-oxido)(2−)]$ **or** metasilicate	$SiO_3^{\bullet-}$ trioxidosilicato($\bullet$1−)
SiO_4			SiO_4^{4-} tetraoxidosilicate(4−) **or** silicate	SiO_4^{4-} tetraoxidosilicato(4−) **or** silicato
Si_2	Si_2 disilicon	$Si_2^{\bullet+}$ disilicon($\bullet$1+)	Si_2^- disilicide($\bullet$1−)	
Si_2H_4	$>SiHSiH_3$ disilane-1,1-diyl $-SiH_2SiH_2-$ disilane-1,2-diyl $=SiHSiH_3$ disilanylidene			
Si_2H_5	$Si_2H_5^{\bullet}$ disilanyl **or** pentahydridodisilicon($Si—Si$)($\bullet$) $-Si_2H_5$ disilanyl	$Si_2H_5^+$ disilanylium	$Si_2H_5^-$ disilanide	$Si_2H_5^-$ disilanido
Si_2H_6	Si_2H_6 disilane (parent hydride name)			Si_2H_6 disilane
Si_2O_7			$Si_2O_7^{6-}$ μ-oxido-bis(trioxidosilicate)(6−) **or** disilicate	$Si_2O_7^{6-}$ μ-oxido-bis▶ (trioxidosilicato)(6−) **or** disilicato
Si_4			Si_4^{4-} tetrasilicide(4−)	
Sn	Sn tin	tin (general) Sn^{2+} tin(2+) Sn^{4+} tin(4+)	stannide	stannido
$SnCl_3$			$SnCl_3^-$ trichloridostannate(1−)	$SnCl_3^-$ trichloridostannato(1−)

Table P7 (*Continued*)

Formula for uncharged atom or group	Name			
	Uncharged atoms or molecules (including zwitterions and radicals) or substituent groups[a]	Cations (including cation radicals) or cationic substituent groups[a]	Anions (including anion radicals) or anionic substituent groups[b]	Ligands[c]
SnH_4	SnH_4 tin tetrahydride **or** stannane (parent hydride name) **or** tetrahydridotin			
Sn_5			Sn_5^{2-} pentastannide(2−)	Sn_5^{2-} pentastannido(2−)
Sr	strontium	strontium	strontide	strontido
T	$^3H^{\bullet} = T^{\bullet}$ tritium(•) **or** monotritium	$^3H^+ = T^+$ tritium(1+) **or** triton	$^3H^- = T^-$ tritide	tritido
T_2	T_2 ditritium	$T_2^{\bullet+}$ ditritium(•1+)		
T_2O	$T_2O = {}^3H_2O$ ditritium oxide **or** ditritidooxygen **or** ($^3H_2$)water			
TcO_4			$TcO_4^{\bullet-}$ tetraoxidotechnetate(1−) TcO_4^{2-} tetraoxidotechnetate(2−)	$TcO_4^{\bullet-}$ tetraoxidotechnetato(1−) TcO_4^{2-} tetraoxidotechnetato2−)
Te	tellurium >Te tellanediyl =Te tellanylidene **or** telluroxo	tellurium	telluride (general) $Te^{\bullet-}$ tellanidyl **or** telluride(•1−) Te^{2-} tellanediide **or** telluride(2−) **or** telluride	tellurido (general) $Te^{\bullet-}$ tellurido(•1−) Te^{2-} tellanediido **or** tellurido(2−) **or** tellurido
TeO_3			$TeO_3^{\bullet-}$ trioxidotellurate(•1−) TeO_3^{2-} trioxidotellurate(2−)	$TeO_3^{\bullet-}$ trioxidotellurato(•1−) TeO_3^{2-} trioxidotellurato(2−)
TeO_4			TeO_4^{2-} tetraoxidotellurate(2−) **or** tellurate	TeO_4^{2-} tetraoxidotellurato(2−) **or** tellurato
TeO_6			TeO_6^{6-} hexaoxidotellurate(6−) **or** orthotellurate	TeO_6^{6-} hexaoxidotellurato(6−) **or** orthotellurato
TiO	TiO titanium(II) oxide	TiO^{2+} oxidotitanium(2+)		
TlH_3	TlH_3 thallium trihydride **or** thallane (parent hydride name) **or** trihydridothallium			
UO_2	UO_2 uranium dioxide	UO_2^+ dioxidouranium(1+) [**not** uranyl(1+)] UO_2^{2+} dioxidouranium(2+) [**not** uranyl(2+)]		

V	vanadium	vanadium	vanadide	vanadido
VO	VO vanadium(II) oxide **or** vanadium mon(o)oxide	VO^{2+} oxidovanadium(2+) (**not** vanadyl)		
VO_2	VO_2 vanadium(IV) oxide **or** vanadium dioxide	VO_2^+ dioxidovanadium(1+)		
W	tungsten	tungsten	tungstide	tungstido
Zn	zinc	zinc	zincide	zincido
ZrO	ZrO zirconium(II) oxide	ZrO^{2+} oxidozirconium(2+)		

[a] Where an element symbol occurs in the first column, the unmodified element name is listed in the second and third columns. The unmodified name is generally used when the element appears as an electropositive constituent in the construction of a stoichiometric name (Chapter 5). Names of homoatomic cations consisting of the element are also constructed using the element name, adding multiplicative prefixes and charge numbers as applicable (Chapter 5). In selected cases, examples of specific cation names such as gold(1+), gold(3+); mercury(2+), dimercury(2+) are given. In such cases, the unmodified element name appears with the qualifier '(general)'.

[b] The 'ide' form of the element name is generally used when the element appears as an electronegative constituent in the construction of a stoichiometric name (Chapter 5), adding multiplicative prefixes and charge numbers as applicable. Examples are given in the Table of names of some specific anions, e.g., chloride(1–), oxide(2–), dioxide(2–). In certain cases, a particular anion is given the 'ide' form itself as an acceptable short name, e.g. arsenide, chloride, oxide. If no specific anions are named, the 'ide' form of the element name is given as the first entry in the fourth column with no further modification, but with the qualifier '(general)'.

[c] Some ligand names must always be placed within enclosing marks to avoid ambiguity. For example, 'dioxido' should be enclosed to distinguish it from two oxido ligands; if combined with a multiplicative prefix the name 'dioxido' should be enclosed because it starts with a multiplicative prefix itself. A ligand name such as 'nitridocarbonato' must always be enclosed when used in a compound name to avoid it being understood as two separate ligand names, 'nitrido' and 'carbonato'. See Chapter 7 for further details.

The charge number on the ligand names given here can be omitted if it is not desired to make any implication regarding the charge. For example, the ligand name [dioxido(•1–)] may be used if one wishes explicitly to consider the ligand to be the species dioxide(•1–), whereas the ligand name (dioxido) carries no such implication.

[d] The ending 'ide' in the names actinide and lanthanide are now taken to denote specifically a negative ion. Actinoid is now the recommended collective name for the elements Ac, Th, Pa, U, Np, Pu, Am, Cm, Bk, Cf, Es, Fm, Md, No, and Lr, and lanthanoid is now the recommended collective name for the elements La, Ce, Pr, Nd, Pm, Sm, Eu, Gd, Tb, Dy, Ho, Er, Tm, Yb, and Lu (see *The Nomenclature of Inorganic Chemistry, Recommendations 2005*).

[e] The name 'phosphorous acid' and the formula H_3PO_3 have been used in the literature for both $P(OH)_3$ and $PHO(OH)_2$.

[f] The two species with the stoichiometry S_2O_5 here are unambiguously named using additive nomenclature. The name 'disulfite' is often cited in the literature for the unsymmetrical structure. However, name disulfite is related to the name 'disulfurous acid', which according to the traditional systematics for oxoacid names should denote the symmetrical form $HOS(O)OS(O)OH$. The name disulfurous acid is no longer acceptable. The name metabisulfite is no longer acceptable in this sense.

APPENDIX

Table P8 Names of mononuclear parent hydrides used in generalised substitutive nomenclature.

Group 13		Group 14		Group 15		Group 16		Group 17	
borane	BH_3	methane	CH_4	azane	NH_3	oxidane	H_2O	fluorane	HF
alumane	AlH_3	silane	SiH_4	phosphane	PH_3	sulfane	H_2S	chlorane	HCl
gallane	GaH_3	germane	GeH_4	arsane	AsH_3	selane	H_2Se	bromane	HBr
indigane	InH_3	stannane	SnH_4	stibane	SbH_3	tellane	H_2Te	iodane	HI
thallane	TlH_3	plumbane	PbH_4	bismuthane	BiH_3	polane	H_2Po	astatane	HAt

Table P9 The names and structures of divalent constitutional units commonly encountered in polymers, based upon the *Compendium of Polymer Terminology and Nomenclature*, RSC Publishing, Cambridge, UK, 2009, the second edition of the so-called Purple Book. This is also reference 1 of Chapter 9 of this book. Most of the names listed here may also be used in organic nomenclature, though some are restricted to use in polymer nomenclature. The names in the left hand column marked with an asterisk (*) may be found in the literature but should not be used in new documents as they are no longer recommended (see references 7 and 16 of Chapter 9). The names in the right-hand column should now be used in polymer nomenclature in preference to those in the left hand column.

Note the use of wavy lines at right angles to the lines indicating the "free" valences. Organic nomenclature practice recommends their employment to indicate free valences attached to ring systems (see Section 6.5, page 71) though they are not always considered necessary for application to linear structures. However, at present polymer chemists never use wavy lines in this manner, in order to avoid confusion with diagrams showing groups bridging between two polymer chains.

Those entries marked † are retained IUPAC names, for different contexts or for groups other than the ones being indicated here, and hence unfortunately may give rise to some confusion. In Chapter 9 (Polymer Nomenclature) the divalent constitutional units are often considered to be in the form -(constitutional unit)-, with 'free valences' attached to different atoms. However, names such as benzylidene and vinylidene refer to constitutional units with both 'free valences' on a single atom as in $CH_2=C=$ and $PhCH=$ being used for bonding them to the same atom of another group, thus forming a double bond.

Name	Structure or synonym
adipoyl	see hexanedioyl
azo	see diazenediyl
azoimino*	see triaz-1-ene-1,3-diyl
azoxy	$-N(O)=N-$ or $-N=N(O)-$
benzoylimino	$C_6H_5CON<$
benzylidene*†	see phenylmethylene
biphenyl-3,5-diyl*†	see 5-phenyl-1,3-phenylene
biphenyl-4,4'-diyl	
butanedioyl	$-COCH_2CH_2CO-$
butane-1,1-diyl*†	see propylmethylene
butane-1,4-diyl	$-[CH_2]_4-$
but-1-ene-1,4-diyl	$-CH=CHCH_2CH_2-$
butylidene*†	see propylmethylene
carbonimidoyl	$-C(=NH)-$
carbonothioyl	$-CS-$
carbonyl	$-CO-$
cyclohexane-1,1-diyl	
cyclohexane-1,4-diyl	
cyclohexylidene*†	see cyclohexane-1,1-diyl
decanedioyl	$-CO[CH_2]_8CO-$
diazenediyl	$-N=N-$
diazoamino*	see triaz-1-ene-1,3-diyl
dimethylmethylene	$(CH_3)_2C<$
dioxy*	see peroxy
diphenylmethylene	$(C_6H_5)_2C<$
disulfanediyl	$-SS-$

232

Table P9 (*Continued*)

Name	Structure or synonym
dithio*	see disulfanediyl
ethanedioyl	see oxalyl
ethane-1,1-diyl*†	see methylmethylene
ethane-1,2-diyl **or** ethylene	$-CH_2CH_2-$
ethanediylidene	$=CHCH=$
ethene-1,1-diyl*†	see methylidenemethylene
ethene-1,2-diyl	$-CH=CH-$
ethylene **or** ethane-1,2-diyl	$-CH_2CH_2-$
ethylidene*†	see methylmethylene
glutaryl	see pentanedioyl
hexamethylene*	see hexane-1,6-diyl
hexanedioyl	$-CO[CH_2]_4CO-$
hexane-1,6-diyl	$-[CH_2]_6-$
hydrazine-1,2-diyl	$-NHNH-$
hydrazo*	see hydrazine-1,2-diyl
hydroxyimino	$HO-N<$
iminio	$-NH_2^+-$
imino	$-NH-$
isophthaloyl	

isopropylidene*†	see dimethylmethylene
malonyl	$-COCH_2CO-$
methanylylidene	$-CH=$
methylene	$-CH_2-$
1-methylethane-1,1-diyl*	see dimethylmethylene
1-methylethylene **or** 1-methyl▶ ethane-1,2-diyl	$-CH(CH_3)CH_2-$
methylidenemethylene	$CH_2=C<$
methylidyne ($-CH=$)*†	see methanylylidene
methylmethylene	$CH_3CH<$
methylylidene*	see methanylylidene
naphthalene-1,8-diyl	

nitrilo	$-N=$
oxalyl	$-COCO-$
oxy	$-O-$
pentamethylene*	see pentane-1,5-diyl
pentanedioyl	$-COCH_2CH_2CH_2CO-$
pentane-1,5-diyl	$-[CH_2]_5-$
peroxy	$-OO-$
1,4-phenylene	

phenylmethylene	$C_6H_5CH<$
5-phenyl-1,3-phenylene	

phthaloyl	

Table P9 (*Continued*)

Name	Structure or synonym
piperidine-1,4-diyl	
propanedioyl	see malonyl
propane-1,3-diyl	$-[CH_2]_3-$
propane-2,2-diyl*†	see dimethylmethylene
propylene*	see 1-methylethylene
propylmethylene	$CH_3CH_2CH_2CH<$
silanediyl	$-SiH_2-$
silylene*†	see silanediyl
succinyl	see butanedioyl
sulfanediyl	$-S-$
sulfinyl	$-S(O)-$
sulfonyl	$-S(O)_2-$
terephthaloyl	
tetramethylene*	see butane-1,4-diyl
thio*†	see sulfanediyl
thiocarbonyl*	see carbonothioyl
triaz-1-ene-1,3-diyl	$-N=N-NH-$
trimethylene*	see propane-1,3-diyl
vinylene*	see ethene-1,2-diyl
vinylidene*	see methylidenemethylene

Table P10 A list of trivial names that may be encountered, together with their more systematic equivalents.

    Names in **bold type-face** are **IUPAC names**. Names in ordinary type-face are **not part of IUPAC nomenclature**. Those preceded by the **sign *** are **not acceptable** for a variety of reasons; they may have been explicitly designated by IUPAC as obsolete, they may be ambiguous because they have been used in the past for more than one structure, or they may have the appearance of systematic names but actually violate particular IUPAC rules for systematic names. The list contains a few class names, INNs, INCI names, and proprietary names (for which see Chapter 12). For specific nomenclature rules, such as the use of parentheses, the reader should consult the appropriate chapters of this book. Where a name in the left-hand column is repeated, this implies that it has been used in the past to name two or more materials. Generally a reader will then have to select the appropriate correct name. As throughout this text, spelling follows UK English practice.

Trivial or common name	Synonym (or explanation)
acetal	**1,1-diethoxyethane**
*acetaldehyde cyanohydrin	**2-hydroxypropanenitrile**
acetic acid	**ethanoic acid**
acetoacetic ester	**ethyl 3-oxobutanoate**
acetoin	**3-hydroxybutan-2-one**
acetone	**propanone**
acetophenone	**1-phenylethan-1-one**
acetylacetone	**pentane-2,4-dione**
acetylcholine	**(2-acetoxyethyl)trimethylammonium**
acetylide	**ethynediide; dicarbide(2–)**
acetylsalicylic acid	**2-acetoxybenzoic acid**
acrolein	**acrylaldehyde; prop-2-enal**
*actinide	**actinoid** (element in the series Ac, Th, ..., No, Lr)
actinide	generic name for a homoatomic anion of actinium
adrenaline	**(*R*)-4-[1-hydroxy-2-(methylamino)ethyl]benzene-1,2-diol**; epinephrine (INN)
alcohol	**ethanol**
aldol	**3-hydroxybutanal**
alum	**aluminium potassium bis(sulfate)—water (1/12)**
alumina	**dialuminium trioxide**
amfetamine (INN)	***rac*-1-phenylpropan-2-amine**

Table P10 *(Continued)*

Trivial or common name	Synonym (or explanation)
ammonia	**azane**
ammonium alum	**aluminium ammonium bis(sulfate)—water (1/12)**
amphetamine	**1-phenylpropan-2-amine**
*amyl alcohol	**pentan-1-ol**
*amyl alcohol	a mixture of isomeric pentyl alcohols
angelic acid	**(Z)-2-methylbut-2-enoic acid**
anhydrite	**calcium sulfate** (anhydrous form)
(o-, m-, or p-)anisic acid	**(2-, 3-, or 4-)methoxybenzoic acid**
(o-, m-, or p-)anisidine	**(2-, 3-, or 4-)methoxyaniline**
antabus	**tetraethylthiuram disulfide**
antimonyl chloride	**antimony(III) chloride oxide**
arachidonic acid	**(5Z,8Z,11Z,14Z)-icosa-5,8,11,14-tetraenoic acid**
*argentic	**silver(III)**
*argentous	**silver(I)**
arsenate	**tetraoxidoarsenate(3–)**
*arsenic	**arsenic(V)**
*arsenic	**diarsenic trioxide; arsenic(III) oxide**
arsenic	name of the element with symbol As
arsenic acid	**trihydridooxidoarsenic**
*arsenic pentoxide	**diarsenic pentaoxide; arsenic(V) oxide**
arsenic trioxide	**diarsenic trioxide; arsenic(III) oxide**
*arsenious	**arsenic(III)**
arsenite	**trioxidoarsenate(3–)**
*arsenous	**arsenic(III)**
arsenous acid	**trihydroxidoarsenic**
*arsine	**arsane**
*arsonium	**arsanium**
*auric	**gold(III)**
*aurous	**gold(I)**
azide	**trinitride(1–)**
azobenzene	**1,2-diphenyldiazene**
bakelite	**poly(formaldehyde-co-phenol)**
*benzene hexachloride	**1,2,3,4,5,6-hexachlorocyclohexane**
benzoin	**2-hydroxy-1,2-diphenylethan-1-one**
betacarotene (INN)	**β,β-carotene**
betaine	**(trimethylammonio)acetate**
biacetyl	**butane-2,3-dione**
*bicarbonate	**hydrogencarbonate; hydroxidodioxidocarbonate(1–)**
*bismuth oxychloride	**bismuth(III) chloride oxide**
*bismuth subacetate	**bismuth(III) acetate oxide**
*bismuthyl chloride	**bismuth(III) chloride oxide**
bisphenol A	**4,4′-(propane-2,2-diyl)diphenol**
*bisulfate	**hydrogensulfate; hydroxidotrioxidosulfate(1–)**
*bisulfite	**hydrogensulfite; hydroxidodioxidosulfate(1–)**
borax	**sodium heptaoxidotetraborate(2–)—water (1/10)**
*borohydride	**tetrahydridoborate(1–); tetrahydridoborate(III)**
British anti-lewisite (acronym BAL)	**2,3-bis(sulfanyl)propan-1-ol; dimercaprol (INN)**
bromoform	**tribromomethane**
*n-butane	**butane**
*sec-butanol	**butan-2-ol**
*tert-butanol	**2-methylpropan-2-ol; tert-butyl alcohol**
*n-butyl	**butyl**
butyric acid	**butanoic acid**
*butyrophenone	**1-phenylbutan-1-one**
caffeine	**1,3,7-trimethyl-7H-purine-2,6(1H,3H)-dione**
calcium cyclamate (INCI)	calcium salt of N-cyclohexylsulfamic acid
calomel	**mercury(I) chloride**
camphor	**bornan-2-one**
cane sugar	**sucrose**
*caproic acid	**hexanoic acid**

Table P10   (*Continued*)

Trivial or common name	Synonym (or explanation)
carbamate	**aminomethanoate**
carbamic acid	**aminomethanoic acid**
*carbamide	**urea**
*carbaminate	**carbamate**
*carbide	**calcium carbide; calcium dicarbide(2−); calcium ethynediide**
carbide	generic name for a homoatomic anion of carbon
*carbolic acid	**phenol**
carborundum	**silicon carbide**
Caro's acid	**peroxysulfuric acid; dioxidanidohydroxidodioxidosulfur**
*catechin	**benzene-1,2-diol**
catechol	**benzene-1,2-diol**
*catechol	**(2R,3S)-2-(3,4-dihydroxyphenyl)chromane-3,5,7-triol**
caustic soda	**sodium hydroxide**
*ceric	**cerium(IV)**
*cerous	**cerium(III)**
cetyl	**hexadecyl**
cetyl alcohol (INCI)	**hexadecan-1-ol**
chloral	**2,2,2-trichloroacetaldehyde**
chloral hydrate	**2,2,2-trichloroethane-1,1-diol**
*chloramine	**chloroazane; chloridodihydridonitrogen**
chloramine-T	**sodium chloro(4-methylbenzenesulfonyl)azanide**
chloroform	**trichloromethane**
chloroprene	**2-chlorobuta-1,3-diene**
chlorosulfonic acid	**chlorosulfuric acid; sulfurochloridic acid; chloridohydroxidodioxidosulfur**
choline	**(2-hydroxyethyl)trimethylammonium**
chromate	**tetraoxidochromate(2−)**
chrome alum	**chromium(III) potassium bis(sulfate)—water (1/12)**
*chromic	**chromium(III)**
*chromic	**chromium(VI)**
chromic acid	**dihydrogen(tetraoxidochromate); dihydroxidodioxidochromium**
*chromous	**chromium(II)**
*chromyl chloride	**chromium dichloride dioxide**
cinnamic acid	**(E)-3-phenylprop-2-enoic acid**
citric acid	**2-hydroxypropane-1,2,3-tricarboxylic acid**
*cobaltic	**cobalt(III)**
*cobaltinitrite	**hexa(nitrito-κN)cobaltate(3−)**
*cobaltous	**cobalt(II)**
cocaine	**methyl (1R,2R,3S,5S)-3-(benzoyloxy)-8-methyl-8-azabicyclo[3.2.1]octane-▶ 2-carboxylate**
colamine	**2-aminoethan-1-ol**
*columbium	**niobium**
coumarin	**2H-chromen-2-one**
creatinin	**2-amino-1-methyl-1,5-dihydro-4H-imidazol-4-one**
cryolite	**sodium hexafluoridoaluminate(3−)**
*cuprammonium	**tetraamminecopper(2+)**
*cupric	**copper(II)**
*cuprous	**copper(I)**
*cyanamide [NCN^{2-}]	**methanebis(iminide); dinitridocarbonate(2−)**
cyanamide [H$_2$NCN]	**aminomethanenitrile; azanecarbonitrile**
cyanate	**azanylidynemethanolate; nitridooxidocarbonate(1−)**
cyanic acid (*cf.* isocyanic acid)	**hydroxidonitridocarbon**
cyanogen	**oxalonitrile; ethanedinitrile**
cyanuric acid (*cf.* isocyanuric acid)	**1,3,5-triazine-2,4,6-triol**
cyclamates	calcium or sodium salts of cyclamic acid
cyclamic acid	**N-cyclohexylsulfamic acid**
dansyl	**5-(dimethylamino)naphthalene-1-sulfonyl**
DDT	acronym from incorrect name for **1,1,1-trichloro-2,2-bis(4-chlorophenyl)ethane**
*decalin	**decahydronaphthalene**

Table P10 (*Continued*)

Trivial or common name	Synonym (or explanation)
dextrose	D-glucose
*diacetyl	biacetyl; **butane-2,3-dione**
diacetylmorphine	heroin
*dichlorodiphenyltrichloroethane	(often represented by DDT, but this is actually a misnomer, see entry for DDT) **1,1,1-trichloro-2,2-bis(4-chlorophenyl)ethane**
dichromate	**μ-oxido-bis(trioxidochromate)(2−)**
dichromic acid	**dihydrogen[μ-oxido-bis(trioxidochromate)]**
*diethanolamine	**2,2′-iminodiethan-1-ol**
diethylacetal	**1,1-diethoxyethane**
diethylene glycol	**2,2′-oxydiethan-1-ol**
diethylenetriamine	***N*-(2-aminoethyl)ethane-1,2-diamine**
*diimide	**diazene**
*diimine	**diazene**
dimercaprol (INN) (*cf.* British anti-lewisite)	***rac*-2,3-bis(sulfanyl)propan-1-ol**
*dimethylglyoxime	**butane-2,3-dione dioxime**
dioxins	derivatives of **oxanthrene** (dibenzo[*b*,*e*][1,4]dioxine)
dioxins	class name for 1,2-, 1,3-, and 1,4-dioxins
*diphenyl	**biphenyl**
*diphosphine	**diphosphane**
ecstasy	**1-(1,3-benzodioxol-5-yl)-*N*-methylpropan-2-amine**
EDTA	**ethylenediaminetetraacetic acid;** **2,2′,2″,2‴-(ethylenedinitrilo)tetraacetic acid**
epichlorohydrin	**(chloromethyl)oxirane**
epinephrine (INN)	**(*R*)-4-[1-hydroxy-2-(methylamino)ethyl]benzene-1,2-diol**
Epsom salt	**magnesium sulfate—water (1/7)**
*ethanolamine	**2-aminoethan-1-ol**
*ethylene	**ethene**
ethylene	**ethane-1,2-diyl**
*ethylene chlorohydrin	**2-chloroethan-1-ol**
*ethylene cyanohydrin	**3-hydroxypropanenitrile**
ethylenediamine	**ethane-1,2-diamine**
ethylene oxide	**oxirane**
*ethyl mercaptan	**ethanethiol**
*ferric	**iron(III)**
*ferricyanide	**hexacyanidoferrate(3−)**
*ferrocyanide	**hexacyanidoferrate(4−)**
ferroin	**(*OC*-6-11)-tris(1,10-phenanthroline-κ*N*¹,κ*N*¹⁰)iron(2+)**
*ferrous	**iron(II)**
flavone	**2-phenyl-4*H*-chromen-4-one**
*fluorosilicate	**hexafluoridosilicate(2−)**
formalin	aqueous solution of **formaldehyde**
freon 12	**dichlorodifluoromethane**
freon 21	**dichlorofluoromethane**
freons	fluorohalohydrocarbons in general
fulminic acid	**formonitrile oxide**
fulvalene	**1,1′-bi(cyclopenta-2,4-dien-1-ylidene)**
furfural	**2-furaldehyde**
furfuraldehyde	**2-furaldehyde**
*furfurol	**2-furaldehyde**
gallic acid	**3,4,5-trihydroxybenzoic acid**
Glauber's salt (or Glauber salt)	**sodium sulfate—water (1/10)**
*glucinium	**beryllium**
*glycerine	**glycerol**
glycocoll	**glycine**
glycol	**ethylene glycol; ethane-1,2-diol**
Graham's salt	**sodium metaphosphate**
guaiacol	**2-methoxyphenol**
gypsum	**calcium sulfate—water (1/2)**
*hahnium (indicating element 105)	**dubnium**

Table P10 (*Continued*)

Trivial or common name	Synonym (or explanation)
*hahnium (indicating element 108)	**hassium**
hemimellitic acid	**benzene-1,2,3-tricarboxylic acid**
heroin	**17-methyl-4,5α-epoxymorphin-7-ene-3,6α-diyl diacetate**
*hexamethylenediamine	**hexane-1,6-diamine**
*hexamethylenetetramine	**1,3,5,7-tetraazaadamantane**
*n-hexane	**hexane**
hippuric acid	**N-benzoylglycine**
hydrazoic acid	**hydrogen azide; hydrido-1κH-trinitrogen(2 N—N)**
hydrochloric acid	aqueous solution of **hydrogen chloride**
*hydronium	**oxonium; oxidanium**
hydroquinone	**benzene-1,4-diol**
*hydrosulfide	**sulfanide; hydrogen(sulfide)(1–)**
*hydroxyl	**hydroxide; oxidanide**
hydroxyl	**oxidanyl; hydridooxygen(•)**
*hyperoxide	**superoxide; dioxide(•1–)**
*hypophosphite	**phosphinate; dihydridodioxidophosphate(1–)**
hypophosphite	**dioxidophosphate(3–)**
imidazoline	dihydroimidazole (with addition of appropriate locants)
iodoform	**triiodomethane**
iron pentacarbonyl	**pentacarbonyliron**
*isoamyl alcohol	**3-methylbutan-1-ol**
*isobutanol	**2-methylpropan-1-ol**
isobutene	**2-methylprop-1-ene**
*isobutylene	**2-methylprop-1-ene**
isocitric acid	**(1R*,2S*)-1-hydroxypropane-1,2,3-tricarboxylic acid**
isocyanic acid (*cf.* cyanic acid)	**iminomethanone**
isocyanuric acid (*cf.* cyanuric acid)	**1,3,5-triazinane-2,4,6-trione**
isooctane	**2,2,4-trimethylpentane**
*isooctanol	**6-methylheptan-1-ol**
*isooctyl	**2-ethylhexyl**
*isooctyl	**6-methylheptyl**
*isopropanol	**propan-2-ol**
isothiocyanic acid	**iminomethanethione**
*joliotium (for element 103)	**lawrencium**
*joliotium (for element 105)	**dubnium**
*kurchatovium	**rutherfordium**
*lanthanide	**lanthanoid** (element in the series La, Ce, ..., Yb, Lu)
lanthanide	generic name for a homoatomic anion of lanthanum
laughing gas	**dinitrogen oxide**
lauric acid	**dodecanoic acid**
lauryl alcohol (INCI)	**dodecan-1-ol**
*lead tetraethyl	**tetraethylplumbane; tetraethyllead**
*levulose	**D-fructose**
lewisite	**dichloro(2-chlorovinyl)arsane**
linoleic acid	**(9Z,12Z)-octadeca-9,12-dienoic acid**
linolenic acid	**(9Z,12Z,15Z)-octadeca-9,12,15-trienoic acid**
linolic acid	**(9Z,12Z)-octadeca-9,12-dienoic acid**
*lithium aluminium hydride	**aluminium lithium hydride; lithium tetrahydridoaluminate(1–)**
luminol	**5-amino-2,3-dihydrophthalazine-1,4-dione**
Maddrell's salt	**sodium metaphosphate**
Magnus' green salt	**tetraammineplatinum(2+) tetrachloridoplatinate(2–)**
malic acid	**2-hydroxysuccinic acid; 2-hydroxybutanedioic acid**
*magnesia	**magnesium carbonate**
*magnesia	**magnesium oxide**
mandelic acid	**hydroxy(phenyl)acetic acid**
*manganese heptoxide	**dimanganese heptaoxide**
*manganic	**manganese(III)**
*manganic	**manganese(VI)**
*manganous	**manganese(II)**
melamine	**1,3,5-triazine-2,4,6-triamine**

Table P10 (*Continued*)

Trivial or common name	Synonym (or explanation)
Meldrum's acid	2,2-dimethyl-1,3-dioxane-4,6-dione
mellitic acid	benzenehexacarboxylic acid
(–)-menthol	(1*R*,2*S*,5*R*)-2-isopropyl-5-methylcyclohexan-1-ol
*mercaptan(s)	thiol(s)
*mercapto	sulfanyl
*mercuric	mercury(II)
*mercurous	mercury(I)
mesitylene	1,3,5-trimethylbenzene
*mesityl oxide	4-methylpent-3-en-2-one
*metabisulfite	pentaoxidodisulfate(2–)
metamfetamine (INN)	(*S*)-*N*-methyl-1-phenylpropan-2-amine
*methyl ethyl ketone (acronym MEK)	butan-2-one; ethyl methyl ketone
*methyl isobutyl ketone (acronym MIBK)	4-methylpentan-2-one; isobutyl methyl ketone
Millon's base	(μ-nitrido-dimercury)(1+) hydroxide
Mohr's salt	diammonium iron bis(sulfate)—water (1/6)
*monoethanolamine (acronym MEA)	2-aminoethan-1-ol
*monopropylene glycol (acronym MPG)	propane-1,2-diol
monosodium glutamate	sodium L-glutamate(1–)
*muriatic acid	hydrochloric acid
mustard gas	bis(2-chloroethyl)sulfane; 2,2'-dichloro-1,1'-sulfanediyldiethane
myristic acid	tetradecanoic acid
*naphthacene	tetracene
α-naphthol	1-naphthol; naphthalen-1-ol
β-naphthol	2-naphthol; naphthalen-2-ol
naphthoresorcinol	naphthalene-1,3-diol
α-naphthylamine	naphthalen-1-amine
neoprene	poly(2-chlorobuta-1,3-diene)
Nessler's reagent	potassium tetraiodidomecurate(2–)
*nickel carbonyl	tetracarbonylnickel
nicotine	3-[(*S*)-1-methylpyrrolidin-2-yl]pyridine
ninhydrin	2,2-dihydroxyindane-1,3-dione
*nitroglycerine	tri-*O*-nitroglycerol; propane-1,2,3-triyl trinitrate
*nitronium	dioxidonitrogen(1+)
*nitroprusside	pentacyanidonitrosylferrate(2–)
nitrosyl	oxidonitrogen(•)
*nitrosyl	oxidonitrogen(1+)
nitrosyl chloride	chloridooxidonitrogen
*nitroxyl	azanone; hydridooxidonitrogen
nitryl	dioxidonitrogen(•)
*nitryl	dioxidonitrogen(1+)
nitryl chloride	chloridodioxidonitrogen
*norbornane	bicyclo[2.2.1]heptane; 8,9,10-trinorbornane
*norbornene	bicyclo[2.2.1]hept-2-ene
*norleucine	2-aminohexanoic acid
*norvaline	2-aminopentanoic acid
oleic acid	(*Z*)-octadec-9-enoic acid
olein	tri-*O*-oleoylglycerol; propane-1,2,3-triyl trioleate
oleoyl	(*Z*)-octadec-9-enoyl
oleyl	(*Z*)-octadec-9-en-1-yl
oxanthrene (*cf.* dioxins)	dibenzo[*b*,*e*][1,4]dioxine
oxazoline	dihydro-1,3-oxazole (with addition of appropriate locants)
*oxine	quinolin-8-ol
*oxine	pyran
oxone	potassium hydrogensulfate—potassium sulfate—▶ potassium hydrogen(peroxysulfate) (1/1/2)
ozone	trioxygen
palmitic acid	hexadecanoic acid
palmitin	tri-*O*-hexadecanoylglycerol; propane-1,2,3-triyl tris(hexadecanoate)
paraformaldehyde	polyformaldehyde; poly(oxymethylene)
paraldehyde	2,4,6-trimethyl-1,3,5-trioxane

239

Table P10 (*Continued*)

Trivial or common name	Synonym (or explanation)
*n-pentane	pentane
peracetic acid	**peroxyacetic acid; ethaneperoxoic acid**
perborate	**di-μ-peroxido-1κO,2κO'-bis(dihydroxidoborate)(2–)**
percarbamide	**hydrogen peroxide—urea (1/1)**; urea peroxide (INCI); carbamide peroxide
permanganate	**tetraoxidomanganate(1–)**
persulfate	**peroxydisulfate; [μ-peroxido-bis(trioxidosulfate)](2–)**
phenolphthalein	**3,3-bis(4-hydroxyphenyl)-2-benzofuran-1(3H)-one**
phenolphthalin	**2-[bis(4-hydroxyphenyl)methyl]benzoic acid**
phosgene	**carbonyl dichloride; dichloridooxidocarbon**
*phosphine	**phosphane**
phosphonium	**phosphanium**
*phosphorus pentoxide	**diphosphorus pentaoxide; phosphorus(V) oxide**
phosphoryl chloride	**phosphorus trichloride oxide**
plaster of Paris	**calcium sulfate—water (2/1)**
*plumbic	**lead(IV)**
*plumbous	**lead(II)**
polyethylene	**polyethene; poly(methylene)**
poly(ethylene glycol)	**poly(ethylene oxide); poly(oxyethylene); poly(oxyethane-1,2-diyl)**
polypropylene	**polypropene; poly(1-methylethylene); poly(1-methylethane-1,2-diyl)**
poly(vinyl alcohol)	**poly(ethenol); poly(1-hydroxyethylene); poly(1-hydroxyethane-1,2-diyl)**
poly(vinyl chloride)	**poly(chloroethene); poly(1-chloroethylene); poly(1-chloroethane-1,2-diyl)**
*polyvinylpyrrolidone	**poly(1-vinylpyrrolidin-2-one); poly[1-(2-oxopyrrolidin-1-yl)ethylene]; poly[1-(2-oxopyrrolidin-1-yl)ethane-1,2-diyl]**
*propargyl	**prop-2-ynyl**
*propylene	**propene**
*propylene	**propane-1,2-diyl**
*propylene	**propane-1,3-diyl**
*propylene glycol	**propane-1,2-diol**
*propylene glycol	**propane-1,3-diol**
*propylene oxide	**2-methyloxirane; 1,2-epoxypropane**
*propylene oxide	**oxetane**
pyrocatechol	**benzene-1,2-diol**
pyrogallol	**benzene-1,2,3-triol**
α-pyrone	**2H-pyran-2-one**
γ-pyrone	**4H-pyran-4-one**
pyrophosphate	**diphosphate; μ-oxido-bis(trioxidophosphate)(4–)**
pyrosulfate	**disulfate; μ-oxido-bis(trioxidosulfate)(2–)**
quicklime	**calcium oxide**
*quinol	**benzene-1,4-diol**
quinone	**cyclohexa-2,5-diene-1,4-dione**
*racemic acid	**(±)-tartaric acid; *rac*-tartaric acid**
red lead	**trilead tetraoxide; dilead(II) lead(IV) oxide**
resorcinol	**benzene-1,3-diol**
resveratrol	**(E)-stilbene-3,4',5-triol**
*rhodanide	**thiocyanate**
Rochelle salt	**potassium sodium L-tartrate**
saccharin (INCI)	**1,1-dioxo-1,2-dihydro-3H-1λ^6,2-benzothiazol-3-one; 1,2-benzothiazol-3(2H)-one 1,1-dioxide**
salicylic acid	**2-hydroxybenzoic acid**
saltpetre	**potassium nitrate**
sarcosine	***N*-methylglycine**
scatole	**3-methylindole**
Seignette salt	**potassium sodium L-tartrate**
semicarbazide	carbamic acid hydrazide
silica	**silicon dioxide**
*silicofluoride	**hexafluoridosilicate(2–)**
silicones	substituted derivatives of poly(oxysilanediyl)
slaked lime	**calcium hydroxide**
soda ash	**sodium carbonate**
sodium cyclamate (INN, INCI)	**sodium *N*-cyclohexylsulfamate**

Table P10 (*Continued*)

Trivial or common name	Synonym (or explanation)
sodium percarbonate	**sodium carbonate—hydrogen peroxide (2/3)**
sorbic acid	**(2*E*,4*E*)-hexa-2,4-dienoic acid**
*sorbitol	**D-glucitol**
speed	**(*S*)-*N*-methyl-1-phenylpropan-2-amine**
squaric acid	**3,4-dihydroxycyclobut-3-ene-1,2-dione**
*stannic	**tin(IV)**
*stannous	**tin(II)**
stearin	**tri-*O*-octadecanoylglycerol; propane-1,2,3-triyl tris(octadecanoate)**
*stibine	**stibane**
sublimate	**mercury(II) chloride**
sulfamic acid	**amidosulfuric acid; amidohydroxidodioxidosulfur**
sulfuryl chloride	**sulfuryl dichloride; sulfur dichloride dioxide; dichloridodioxidosulfur**
*sulphate	**sulfate**
*sulphite	**sulfite**
*sulphur	**sulfur**
tartar emetic	**dipotassium [di-μ-tartrato(4–)-diantimonate(III)]—water (1/3); dipotassium {bis[μ-(2,3-dioxidobutanedioato)]▶ diantimonate}(2–)—water (1/3)**
taurine	**2-aminoethane-1-sulfonic acid**
*tetrahydroborate	**tetrahydridoborate(1–)**
*tetralin	**1,2,3,4-tetrahydronaphthalene**
*thallic	**thallium(III)**
*thallous	**thallium(I)**
*thioacetone	**propanethione**
thiocyanate	**nitridosulfidocarbonate(1–); azanylidynemethanethiolate**
thiocyanic acid	**sulfanylmethanenitrile**
thiocyanogen	**disulfanedicarbonitrile**
*thioglycolic acid	**hydroxythioacetic acid**
*thioglycolic acid	**sulfanylacetic acid**
thionyl chloride	**thionyl dichloride; sulfur dichloride oxide; dichloridooxidosulfur**
*thiophenol	**benzenethiol**
thiosulfate	**trioxidosulfidosulfate(2–)**
tiglic acid	**(*E*)-2-methylbut-2-enoic acid**
titania	**titanium dioxide**
*titanic	**titanium(IV)**
*titanous	**titanium(III)**
TNT	**2,4,6-trinitrotoluene**
*p-toluenesulfonic acid	**4-methylbenzenesulfonic acid**
as-triazine	**1,2,4-triazine**
s-triazine	**1,3,5-triazine**
triclosan (INN)	**5-chloro-2-(2,4-dichlorophenoxy)phenol**
*triethanolamine	**2,2′,2″-nitrilotriethan-1-ol**
*triethylenediamine	**1,4-diazabicyclo[2.2.2]octane**
triflate	**trifluoromethanesulfonate**
*triflic acid	**trifluoromethanesulfonic acid**
triglycerides	**tri-*O*-acylglycerols**
*trimethylenediamine	**propane-1,3-diamine**
triolein	**tri-*O*-oleoylglycerol; propane-1,2,3-triyl trioleate**
tripalmitin	**tri-*O*-hexadecanoylglycerol; propane-1,2,3-triyl tris(hexadecanoate)**
*tripolyphosphate	**triphosphate**
tristearin	**tri-*O*-octadecanoylglycerol; propane-1,2,3-triyl tris(octadecanoate)**
trotyl	**2,4,6-trinitrotoluene**
*uranyl	**dioxidouranium(V)**
*uranyl	**dioxidouranium(VI)**
*urethane	**ethyl carbamate; carbamic acid ethyl ester** (*NB: not* ethylcarbamate)
urethanes	class name for substituted carbamic acid esters
uric acid	**1*H*-purine-2,6,8-triol**
urotropin	**1,3,5,7-tetraazaadamantane**
valeric acid	**pentanoic acid**
*vanadic	**vanadium(III)**

Table P10 (*Continued*)

Trivial or common name	Synonym (or explanation)
*vanadic	**vanadium(V)**
*vanadium pentoxide	**divanadium pentaoxide**
*vanadous	**vanadium(II)**
*vanadyl	**oxidovanadium(IV)**
vanillic acid	**4-hydroxy-3-methoxybenzoic acid**
vanillin	**4-hydroxy-3-methoxybenzaldehyde**
veratrole	**1,2-dimethoxybenzene**
white arsenic	**arsenic(III) oxide**
white lead	**lead carbonate hydroxide**
*wolfram	**tungsten** (name for English language texts, wolfram is used in some other languages)
xanthogenates	**carbonodithioic acid *O*-esters**
Zeise's salt	**potassium trichlorido(η^2-ethene)platinate(1−)—water (1/1)**

Index

Element names, parent hydride names and systematic names derived using any of the nomenclature systems described in this book are, with very few exceptions, not included explicitly in this index. If a name or term is referred to in several places in the book, the most informative references appear in **bold type**, and some of the less informative places are not cited in the index. Endings and suffixes are represented using a hyphen in the usual fashion, e.g. -ol, and are indexed at the place where they would appear ignoring the hyphen. Names of compounds or groups not included in the index may be found in Tables P7 (p. 205), P9 (p. 232) and P10 (p. 234).

INDEX

INDEX

INDEX